Potbellied Pig Behavior and Training

POTBELLIED PIG BEHAVIOR AND TRAINING

A Complete Guide for Solving
Behavioral Problems
in Vietnamese
Potbellied Pigs

by PRISCILLA VALENTINE

All photograghed pigs contained herein are pets of Steve and Priscilla Valentine.

Permission has been obtained for reprinting photos included in this book.

Published by Luminary Media Group, an imprint of Pine Orchard, Inc.
Visit us on the internet at www.pineorchard.com

Printed in Hong Kong.

1st edition ISBN 1-930580-03-7	*2000 (4 color pages.)*
2nd edition ISBN 1-930580-74-6	*2005 (Revised with 16 color pages.)*
	2007 (Revised with 24 color pages.)
	2009

EAN (ISBN-13): 978-1930580749
Library of Congress Card Number: 2005921142

DEDICATION

I dedicate this book to the most incredible and sensitive "person" on Earth, "Nellie." Nellie inspired me, changed my life, and took my family in a different direction. I am in awe of her intelligence. I love her with my heart and soul.

There will never be another Nellie.

Nellie — a firecracker at eight years of age.

CONTENTS

Preface

After Vietnamese potbellied pigs became very popular as pets, it soon became apparent that these critters with the pudgy physique, snouty smiles, and pencil tails had complex, inquisitive personalities. Dog-training methods didn't work, and some pigs became spoiled and headstrong.

Many owners had bonded to their pig deeply. Suddenly, they had a pig that refused to go outdoors or had an attitude toward their houseguests. Pigs were making people's houses their own private castles and would lash out at anyone who dared enter their domain. Porkers were displaying signs of boredom and taking it out on the refrigerator and the knocked-over wastepaper baskets. Landscaped yards were turned into minefields. Wallboards were being chewed upon. The family was being woken at 4:00 in the morning by a ravenous, screaming porker.

But potbellies did not really make bad pets; humans were just understandably bad caretakers. Owners had no idea WHY their pets were behaving this way, let alone how to control their pigs. There was not one single book available that was devoted to potbellied pig training and behavior.

Unfortunately, the pigs suffered because of this. They were often abandoned in shelters as mystified, tearful owners said their good-byes.

Most of what I have learned about a potbellied pig's mind has come from living with pigs in our home for a decade. I have a tremendous passion for these exceptional animals and spend the better part of most days either cuddling or training our bristly family members. I am

totally devoted to this clever species. For me, it was always simple to potty train, obedience train, and halter train a piglet. It is not a job for me; it is a pure joy. We have a mutual admiration for each other.

I feel obliged to pass on my experience in the hope that fewer pigs will be discarded because they seem difficult to handle. Potbellies are incredibly intelligent animals that are adaptable. A single kernel of popcorn can easily motivate them, and there is no reason why anyone cannot train their pig to be a welcome addition to the family. Pigs did not ask to be our pets. They have no choice. It is up to us to teach them how to be GOOD pets in our realm.

Introduction

A little more than a few years ago, no one would have considered having a miniature pig as a companion pet or ANY pig as a household pet. But pigs have been domesticated for thousands of years and the diminutive potbellied pig has the potential to make a great companion pet for some people.

Those people who appreciate the uniqueness, sensitivity, and intelligence of this clever species have come to adore them! Besides being loved, they are also sometimes misunderstood.

It has been said that having a potbelly as a pet is like taking care of a two-year-old for 15 years. Certainly, this is true. Potbellies are mischievous, curious, food-crazy, and can even be convoluted and manipulative. The pigs use most of their cleverness to get food, and it is always a challenge to keep them out of it.

Speaking of intelligence, some zoologists rank the pigs as the most intelligent domesticated animal as well as the brainiest four-legged animals. The University of Kentucky did a five-year study of the intelligence of pigs versus dogs, and the pigs came up on top. Areas such as problem solving and memory were compared.

Intelligent animals need stimulation and challenges. Training a pig can be a very rewarding experience. Pigs not only love food, they live for food. Positive reinforcement can turn a shy pig into a trick champion or just an obedient, beloved pet.

Porkers tend to be a bit lazy and they can be very demanding. Certainly the term "pig-headed" is an accurate one! Pigs are stubborn. Push them to the right,

and they will certainly move to the left. Pigs take nothing for granted. They are foragers by nature and take great pride in uprooting things simply to find out what's under them. This can include dining-room chairs and even the carpet!

Pigs are not the pets for everyone. They are special pets for special people. A pig owner must have a sense of humor! They are not dogs in pig costumes. You must earn a pig's trust, whereas dogs will love you almost immediately. A pig must be allowed to be a pig. Pigs' personalities are very different from dogs' personalities, and they need special training that is tailored to them specifically. Pigs are not motivated to do actions just for the thought of pleasing you, the owner. But, fortunately, pigs love treats and stimulation and something to keep their large brains occupied. That's where training comes in!

Because of pigs' high intelligence, they can often take advantage of being spoiled and act territorial. Pigs need to know exactly how they fit in the family social order. They need to be respected; in turn, they will respect their caretakers. Potbellies demand to be treated fairly, or they won't respect you. And if they don't respect you, they won't like you.

Training a pig is similar to training a child. Their remarkable memory will make working with them a pleasure. A pig may remember how to do a trick that he was trained to do five years ago but has never done since. A pig will also never forget it if you give him a treat from the dining-room table just once! Pigs have the ability to problem-solve and can learn quickly. They love tricks because it's a game for them to perform and be rewarded.

Being fun-loving animals, pigs live only for today. As the most intelligent of all domesticated animals, potbellies have much to offer us. But it is up to you to meet your pig's needs, understand why he behaves the way he does, and take responsibility for shaping his personality so that you can live in mutual harmony and bliss.

CHAPTER 1

VIETNAMESE POTBELLIED PIG BEHAVIOR

A Language of Their Own!

Anyone who has ever been introduced to a potbellied pig is amazed by the variety of bizarre and intriguing sounds he can make.

Communication can range from a gentle, content, low "oofing" sound to a shrill, ear-piercing scream. It is said that pigs make 14 distinct communication noises, each one meaning something special. These sounds include squealing, chomping, grunting, barking, "woofing," lip smacking, "ah ah ah-ing," snorting, and many more noises that are too complex to even describe. Many of the sounds are nasal.

When it's time to let the milk down, mother pigs softy call their hungry babies over (with a distinctive sound) to nurse. When reclining mother pigs are nursing their young, they make a lovely, soft, rhythmic, grunting noise as the babies pulsate at her nipples, eagerly sucking as hard as they can.

Pet pigs that are "bunk mates" make clucking noises at each other, passing the time and sharing stories.

All potbelly owners know the contented, staccato sound of grunting when they are treating their little darlings to a belly-rub massage. A housepig that wants attention will whine.

When a pig is suddenly frightened, he barks a startling call of warning to the rest of the herd, communicating to them the threat of impending danger. New potbellied pig owners are always surprised by the high pitched squeals of a pigger that does not appreciate being restrained. And it doesn't take long to recognize the appreciative "ruf, ruf, ruf" as a porker buries his snout deep in the dinner bowl!

Unlike dogs, threatened potbellies rarely make noises before they defend themselves. They often have a stationary look in their eyes, followed by a quick lunge. It can be very difficult to anticipate. Pigs tend to be silent when they are upset.

When potbellied pig owners pick their pigs up from the kennel after their vacation, they are often greeted by a dog-like, enthusiastic bark that tells them they have been sorely missed!

Piglet Recreation and Learning

Young pigs love to play. Littermates that are only a couple weeks old insist on constantly nudging each other and playing hide and seek and king of the mountain (using the sow as the mountain!). It is easy to kill an entire afternoon watching energetic snub-nosed piglets frolic. Nothing is more adorable than a healthy, playful piglet! There seems to be no limit to their energy as they jump over each other, fight over a morsel of food, or constantly

climb over each other, creating a "pig pile."

They take turns rubbing each other's bellies and are insistent upon pestering tired, old mom by playing with her ears.

Piglets in the same litter behave almost as one single pig. If one piglet is frightened and flees, they all will follow. If one baby pig looks to the left, they all will. Their young lives are a game of "follow the leader." There is hardly any apparent individualism at this young age.

The mother will discipline the pigs with a sharp nose shove or a grunt of disapproval. She will also impart to them the dangers and pitfalls of life while the pigs are with her for the first 8 weeks.

As the pigs get a little older (about 6 months), they do slow down a bit, but they still like to race across pastures or family rooms and will suddenly do a 180-degree turn, flipping their hind quarters in the air! Some amused pet owners affectionately call this "the rodeo dance."

After the pigsters turn one year old, most of their desire to play comes in the form of outsmarting their owners, teasing the dog in order to steal a meal, or getting into mischief.

Pigs enjoy scratching themselves on anything not nailed down. Unlike other pets, they would rather be scratched or rubbed than petted. Most of all, they love their pendulous bellies being given a soft massage.

There is even a belly-rubbing contest at most pig shows. A stopwatch is used, and the first porker to roll over on his side in pleasure wins the blue ribbon!

Foraging and Rooting

In the wild, pigs get the bulk of their daily meals by foraging and exploring. Even though many types of pigs have been domesticated for thousand of years and fed grain by farmers, all pigs still adore foraging for food. Pigs are omnivores and delight in turning over rocks, leaves, branches, and the soil in their eternal quest for goodies.

Their snout is like a sharp digging tool, and it only takes a few seconds for any kind of pig to create a deep hole in the ground. Pigs enjoy nature to the fullest. Most of what interests a pig during his lifetime is at ground level.

Potbellies, as pets, may be very well fed but still desire to root. Some owners confuse this foraging with hunger. Pigs like to root and create havoc because the exploration stimulates them and gives them something to do. It's a game and sport to them, like the sport of fishing is to humans. It occupies the potbelly's time, and it is a challenge for the pig to find something tasty all by himself.

Potbellied pigs always appear to be hungry, even after finishing a large meal. It is no wonder that they have earned the name "pig."

Body Language

Pigs display many of their emotions through body language and stature. In a group of pigs, often there are virtually no sounds being made. But, by careful and persistent observation, you can find out who the leader is by the postures of the pigs. The lesser pigs will walk

Pigs are curious animals that love to explore.

Piglets imitate one another's actions and react as a group.

Pigs hate cold weather and often need some “inspiration” to tolerate it.

lower to the ground and give the right of way to the alpha pig, who may be posturing, standing tall, with his "hackles" (mane) up.

If a porker is upset with another pig, he may swing his head as a warning. This is called "sideswiping" and usually indicates a warning. Wild pigs and potbellied boars have long tusks, which protrude from the side of their mouths, making this gesture a very threatening one. (Neutered pet pigs will grow tusks also, but they will be shorter.) Agitated pigs will have their ears in a down position.

Pigs seem to use their most distinctive feature, their snout, for almost everything. They will nudge each other to show both anger and affection. If the pig is unhappy, a hard nudge can be a precursor to nipping.

With baby piglets, nudging can be a sign of affection. Piglets nudge their mother, almost constantly, hoping to stimulate milk flow. Each piglet has his own nipple that he nurses on. They are usually not far away from it, or are actively pushing on it in the hope that they will get lucky. If mom isn't around, they take turns practicing on each other. Their tiny noses push on one pig's belly as he lies down, enjoying the massage.

When pet potbellied pigs want attention, the snout comes in handy again. A firm push to the ankles of the unsuspecting owner is administered with the purpose of reminding the cook that dinner is two minutes late! It is used for clever manipulation and as a way to communicate. Pigs will also use their noses to make you aware that you have forgotten them for a few minutes. It is used to get attention.

The pig's posture can tell you what he is feeling. Sick

or depressed pigs will walk with a hunkered-over stance. Their hind legs will be carried slightly beneath them as they walk. They do not stand as tall and they keep their head lower to the ground.

A very happy, healthy pig will wag his straight tail so fast it looks like a propeller. As the potbellies age, it becomes more of a constant swish-swish motion from side-to-side.

Fighting pigs will swing their heads and fight side-to-side in a circular motion, like an impending tornado.

There will be little or no sound coming from the sparring opponents. Each pig will attempt to shove the other, making him retreat, so that the fight can end. There may be some snapping or grabbing, especially to the vulnerable ears. Their heads will be down, and their bodies take on a "C" shape as they struggle. The loser will slink away after he is forced to retreat. The winner will acquire social dominance. The pecking order will be set, and the pigs will all be docile and content, as usual.

Clean or Dirty?

In literature, as well as society, pigs have been portrayed as filthy and smelly. There is a negative connotation when a human is called a "pig."

Actually, pigs have meticulous toilet habits, and naturally defecate and urinate in one defined area. Any animal can only be as clean as his environment. A farm pig that is kept in a muddy stall with piles of feces in it will naturally be smelly and dirty. He has no choice but to be that way.

Pigs are not smelly in their natural environment. They do crave a good wallow, but this is because the pig has virtually no sweat glands to keep him cool. The mud acts as a natural skin conditioner (women pay good money for mudpack facials) and as a bug repellent and sunscreen.

If the porksters have no access to mud, the cool, dug-up soil will do. In warm weather, pigs will seek out cool, refreshing water, any way they can, to splash in. Even a knocked-over watering dish can provide a refreshing mini-bath.

Except for boars, pigs have no discernable body odor. Many pet pig owners do not give their piggies baths. They simply brush them once a week in order to loosen dry skin.

Pigs and Romance

When a female pig is in heat (estrus), the boar will ardently pursue her. He will use his snout, fervently pressing the tummy of his chosen, in order to get her "in the mood." When the time comes, the gilt will just stand there, almost like she's frozen.

Ultimately, she will be mounted for roughly five to eight minutes. Afterwards, the pigs will stand, side by side, looking at each other just like young lovers.

Often, they will fall asleep together. But the romance is short-lived. As soon as the female is out of heat, the passionate boar will be off to greener pastures.

Unspayed female pet pigs can act obnoxious during the three-day heat cycle. They can be overly assertive, jumping on people's legs, holding on to pant legs, or nipping. Or they can act spacey, whiney, and forget their restroom habits. Among piggy owners, this is often referred to as "piggy PMS."

Boars must be the most macho of all animal species! They will mount any object and are virtually untrainable. They constantly seek out females and are almost impossible to contain behind a fence.

The clever, amorous animals have been known to dig out of or climb over just about any type of fence in pursuit of love.

They have a strong, unpleasant body odor and will go to great lengths to patrol their territory by pacing the fence. They are incorrigible, willful critters that will fall in love with whatever animal is available or perhaps even with a bucket or a discarded tire.

A boar will not take "NO" for an answer, and no sane person would keep a potbellied boar as a pet. A neutered boar is called a "barrow" and can make an excellent companion.

Pigs and Affection

Pigs do not always show affection toward humans as other classic pets do. They are subtle and often hard to read at first. Most pigs like to be cuddled, but not all do.

A potbelly may not seem to care for his owner that much, outwardly, but will go into a deep depression if he loses him. This can even claim the life of a rejected piggy!

Sometimes, pigs show affection by wagging their tails and following their owner from room to room. It is not unusual for a potbelly to punish his owner by pouting if he is taken to a kennel for a few weeks. A pig can be very, very bonded to his caretaker, almost like a child is

to a parent.

Potbellies first show most of their affection toward their caretakers by trusting them. It is a great honor to be trusted by the often fearful potbellied pig. Potbellies love to be scratched and to sleep in their owner's laps.

Most pet owners spend a lot of time on the floor, rubbing and enjoying their piggies.

Nesting

All pigs love to nest. Once the materials are collected for the bedding, a great ritual is made of distributing pieces of leaves, debris, straw, or blankets. The pigs love to utilize their snouts in making a cocoon-like enclosure of tightly pulled blankets that may completely cover their body. Between using their hooves and snouts, they can create tension on a blanket, making a very efficient and warm body wrap.

It gives new meaning to the phrase "pig in a blanket."

Pigs seem to enjoy nesting and bed making and focus in on it intensely. Pet pigs love to sleep with their owners, but often gravitate toward the very bottom of the bed, providing a welcome foot warmer to their bedmate!

When a female pig is ready to farrow (give birth), she will spend hours making a nest (whether she needs to or not). This is a ritual she must go through in most cases. It is very important that a farrowing potbelly be provided with plenty of material to work with.

Potbellied pigs love to sleep, and one has to wonder if they would ever rise if it weren't for food being served. In the winter, a pig may go to bed as early as 4:00 in the afternoon. In the summer, he may forage until sundown, which can be quite late.

CHAPTER 2

CHOOSING THE RIGHT PET PIG

Vietnamese potbellied pigs are the most popular types of what is commonly referred to as "miniature pigs."

Keith Connell of Oshawa, Ontario, imported the first 18 of the breed in 1985. The pigs are indeed small when compared to commercial hogs that can weigh 800 pounds at maturity. The average weight of a registered, purebred potbelly is about 90 to150 pounds. Some pigs weigh more, some less. Potbellies continue growing until the age of three, or even older, and can be difficult to motivate to move if allowed to be overweight. Both nutrition and the genetics of the relatives of the pig will affect his ultimate size potential. Because of their potential to grow rather large, it is even more important that the pigs be trained to be obedient.

Potbellied pigs are more challenging to take care of than either dogs or cats. Pigs need hoof care, tusk care, skin care, special diets, and something to keep them entertained. A good veterinarian to take care of them is often hard to find. Their diets must be carefully monitored, the house must be pig-proofed, and the potbellies can become spoiled and demanding.

However, if people are motivated and willing to educate themselves before buying a pig, these problems can be prevented. Owning a potbelly can be very enriching to a person's life.

Male or Female

Potbellies, in general, do not make good pets unless they are spayed or neutered. The intact males (boars) are aggressive, hormone-driven, and emit a musky, rank "cologne." No sane person would want a boar as a housepet. They are obnoxious and untractable, and will mount anything within reach.

Females generally do not make good pets either. The intact female can become demanding and whiney during heat (estrus), and she will often forget her housebreaking habits during this three- to five-day period. They are irritable and unpredictable.

Either an altered male (barrow) or an altered female has the potential to be a good companion. Females, in general, have a more serious personality and better focus. They are a little more trainable at the advanced feats. Altered males generally have a more happy-go-lucky personality and a little better sense of humor. Pig devotees can't decide among themselves which sex makes a better pet. It is probably more contingent upon the particular individual personality of the pig than the gender of the pet.

Personalities

Personalities of individual pigs vary widely. Some pigs are bolder, friendlier, and more outgoing. Others seem more sensitive and take longer to trust their owner. Some pigs love to cuddle; others are a little more aloof. If a person owns ten pigs, each one will have a different personality.

It is always good to select a baby pig from a reputable breeder. Buying from a breeder who has purebred,

Piglets should be used to being touched all over before they are harness trained.

Pigs need to spend time outdoors to satisfy their foraging instinct.

registered stock will insure that the bloodlines were not mixed with commercial hog lines and that the pigs were not inbred. A good breeder selects her mating pigs with emphasis on personality as well as desirable physical traits. In pigs, a docile, tractable personality can be inherited to some extent. Or if the birthing sow has been taught to enjoy and trust humans, she will not teach the babies to fear them as much. The breeder should socialize and potty-train the piglet in her own home before selling it as a pet. This puts less stress on the pig and shapes his personality for the future.

Weaning Woes

If a piglet is weaned too early, it could affect his ultimate personality. Pigs that are weaned too early miss the training that the sow and his littermates could have given him. As the piglets get older, they learn to play and fight with each other. Through these actions, they learn appropriate behavior from the sow and the social interaction between brothers and sisters. A piglet that has been weaned before five weeks and taken from his family may not have the social skills that are required to be a good companion. Early-weaned piglets are often demanding and assertive. They may be nippers too. A breeder should not sell a piglet before it is six to seven weeks of age.

Young or Old?

A baby pig is a handful unless the breeder has properly socialized it. Even so, it will take a long time for the piglet to trust you. Adopting an older pig from a pig

sanctuary or rescue is an option for many. Not only does it provide a home for an unfortunate porker, but it also allows the new owner to pick between the pigs in order to find the personality he desires. The pig will already be potty-trained and used to people and other pet animals. He will probably trust you and appreciate a new home.

Some people prefer a baby pig to bond to and "shape." They want a pig they can hold in their arms. It's up to the individual as to what age pig would be the best to adopt.

Making the Decision

The new piglet should be genetically sound and healthy-looking (robust, roly-poly, playful, and alert). He should come from a reputable breeder that properly socializes and weans. If you are able to select a pig yourself from a litter of 8-week-old piglets, look for the piglet that seems comfortable with you and does not retreat as much as the others. Watch which pigs' ears perk up when you talk to them. Ask the breeder if you can get in the pen. Once you do, stretch your hand out and feed the little piglets. Try to see the parents and appraise the parents' sizes. (Remember, a pig keeps growing until it is 3 years old.) It is roughly a 15-year commitment to own a potbellied pig.

CHAPTER 3

BRINGING HOME THE NEW PIG

Piglets

Taking care of a piglet is similar to taking care of a child. Piglets need supervision, confinement, training, kindness, and routine. They are a joy to spend time with, and having a piglet in the house can be a rewarding experience.

1. SUPERVISION — The piglet must always be monitored when he is out of his confined area. This way, poor behavior or toilet habits can be corrected.

2. CONFINEMENT — The piglet should be put in a small area so that he will feel secure and not frightened. Pigs are nesters and don't like to be out in the open until they are used to their surroundings. A laundry room, crate, playpen, etc. works nicely. The pig will learn good potty habits if confined to a small area at first.

3. TRAINING — The piglet should be called for meals and learn the words "come," "good pig," "treat," and the sound of his name right away. He should be praised for

good behavior and reprimanded for undesirable behavior in a low, guttural tone of voice.

4. KINDNESS — Being sensitive to your pig's needs and not causing him to become distressed will start the relationship out well. Spend a lot of time with the frightened piglet the first few days.

5. ROUTINE — A structured environment will make your pig start to feel at home. Naps, training sessions, meals, and potty breaks should be at the same time every day.

Picking Up the Pig at the Airport or Breeder's Home

When picking up your new piglet from the airport or bringing an older pig home for the first time, it's important that you start out on the right "hoof," so to speak. Pigs have excellent memories . . . they are very sensitive and aware animals. Your first impression will have a big impact on whether the pig sees you as a friend or someone to fear.

Potbellies have a strong sense of self-preservation. They are not predators like dogs or cats. Their first priority is the will to survive. They do not have jaws that enable them to defend themselves well. They cannot jump or run efficiently, except when they are very young. They must always be aware of immediate danger. For this reason, they are apprehensive and cautious. They tune into the sense of motion above their heads in order to determine if a predator is about to grab them and carry them away. They must have the use of their feet in case they must run to hide from danger. Running is virtually the porkers only defense. Therefore, all four feet must be on the ground with solid footing. A smart pig will always be cautious throughout his entire life. He will be careful and somewhat reticent if picked up. A mother pig never picks up her babies like dogs and cats do. Only predators lift up pigs in the air in order to make a meal out of them!

A pig knows that other animals prey upon him. Baby pigs are the most vulnerable. When you first get your baby pig, he probably will be very frightened and attempt to run from you. It is very important that the breeder work with the baby piglets very early in life. The best breeders will start socializing the piglets the very day that they are born. Much of the piglet's fear of humans is inborn, but some of it is learned from the sow. As they grow older, the mother pig will teach the piglets to protect themselves from harm.

Surprisingly, a newborn piglet will rarely scream if picked up the day he is born. As the days pass, he will object more and more to being lifted. For this reason, a good breeder starts gently picking up the piglets from day one. But no matter how well the breeder has practiced this, a smart piglet will still scream when you, a stranger, attempt to lift him. This is because the piglet has lost all control when he is raised up. His only defense against predators, the ability to run, has been lost. His screaming is intended to startle you, and is very piercing and penetrating. It is said to be louder than a Concord jet! The pig is hoping this will shock you, forcing you to drop him, so he can run away to hide and be safe.

Potbellies also fear any type of restraint (such as halters), being held down, or being grabbed. Again, this

is because a pig is vulnerable to harm when he is restrained. As a pig learns to trust his new owner, he will gradually tolerate being picked up and restrained. But this must be a gradual process. If the pig screams, you are going too fast.

Because the potbellies often lead a semi-wild existence in Vietnam, their survival instincts are more apparent than in other pets, such as dogs and cats. In order to make a good pet out of a miniature pig, it is imperative that you gain the piglet's trust from the very moment you set your eyes on him at the airport or the breeder's home. Talking in a soothing, low voice will help and attempting to do nothing that will distress the pig will aid in the socialization. Potbellied pigs do not handle stress well. You must be aware of your actions and how the pig sees these actions. If the pig sees his new owner's actions as threatening, it will set back the relationship.

Patience is very important when socializing a new pig. Going slowly will allow the pig to realize that his new owner is not his enemy but his friend. Being a leader to the pig will make him feel secure.

The Transition

Picking up the new piglet or older pig in a dog kennel or airline crate is an excellent idea because the pig will feel secure in the crate, while you have time to introduce yourself. Pigs should never travel in a car without being crated. The motion of the car frightens the pig, and he can get injured very easily if the vehicle stops suddenly. Talking to the pig when you first see him and teaching him his name is an excellent start. A few kernels of pig feed, popcorn, or some grapes will motivate the pig to be interested in developing the relationship. Saying the pig's new name and the word "come" will start the training program off nicely while the pig is still in the crate. He will approach you while in the crate because he knows he is safe (the crate door separates you). This is an excellent opportunity to establish yourself as both the pig's leader and friend without scaring him.

When you get home with the pig, put him in a secluded, noise-free area of the home to allow him to adapt to the surroundings. Let him be alone for a half-hour, still in the crate, then return, teaching him to come to the front of the crate and get paid for coming when his name is called. When you wish to remove him from the crate, do it in an area that is closed off (confined). The pig will probably want to flee.

Call his name, and gently and patiently coax him from the crate. If he will not budge, get on the floor and lift him up, just a few inches off the ground, and set him in a closed area where he will have good footing. Piglets detest slippery floors because they lose their ability to run away if there is trouble. Making the pig feel stable and secure is a top priority. A smaller room for these first days, such as a laundry room, is ideal.

Have some water, food, toys, and blankets ready for your bristly new arrival.

A Look into the Litterbox

Setting up a good litterbox situation before the piglet arrives is imperative. Because the frightened baby pig may run away if he is placed outside, a litterbox should be used at first. Good litterbox training from the get-go

will ensure good elimination patterns in the future. Supervision, confinement, and routine will help the potbelly establish good potty habits.

Pigs naturally have fastidiously clean restroom habits. When a mother pig gives birth, the newborns, on the very day they are born, will totter off to the far corner of the farrowing pen to defecate in the spot their mother uses. Pigs use their keen sense of smell to lead them to the proper area. And pigs are creatures that like routine and predictability. Porkers will almost never eliminate near their eating or sleeping areas. These facts must be taken into account when the litterbox is initially set up.

Potbellied pigs have very short legs and are not the most athletic animals in the world. It's important that the pig is encouraged to use the box, and that the box is easy to enter and exit. A low, cat litterbox with an opening cut in the side for the pig to enter will do fine. Of course, the box should be spacious and allow the pig to turn around in easily. A box with high sides will make the pig reticent to enter. Pigs are leery of being trapped by predators. A smart pig will not enter a tall box where he is vulnerable. The sides should be low enough so that the pig does not have to jump in and out.

The correct litter material is very important. The common, clay, cat-type products are not appropriate for several reasons. The pig may ingest some of the litter, which can cause obstructions in the digestive tract. This type of litter is often dusty and pigs are vulnerable to respiratory problems from inhaling dust particles.

Cedar shavings have been known to cause toxic reactions in some pigs.

PINE SHAVINGS make the ideal litter for piggers. They are natural, non-toxic, and best of all, seem to draw the pigs into the box. They are available at most pet stores. At pig shows with pine-shaving arenas, potbellies will go out of their way to do their business in the arena. One whiff of the pine scent and the pig is motivated!

The litterbox should be placed away from the pig's eating and sleeping area. But it should not be placed out of sight or TOO far away. A corner of the small room where the pig is being confined is ideal. For training purposes, keeping the piglet or new pig in a very small room will help him learn his routine. For the first week, unpleasantries should be kept to a minimum. It's important to convey to the newcomer that he is safe. Bathing the pig should be postponed, if possible. The amount of handling that it requires will set the socialization back. If the pig gets soiled from traveling, a light brushing will help. Other pets, as well as tiny children, should be kept away from the pig. Try not to pick him up, unless it is necessary, for a few days. It is crucial to establish a calm environment for the new piggy.

To introduce your new pig to the litterbox, it would be ideal if a dropping left over from his carrier could be placed in it. The smell will be a signal to the pig that the box is the proper place to potty. If the pig does urinate outside the box, cleaning it up with a paper towel and placing the paper towel inside the box will encourage him to enter the box the next time. Leading the pig to the box or very carefully placing the pig in it will make him aware of it.

Try not to either punish or reward the pig with food for his elimination habits. Praise is fine. Pigs naturally

Older pigs can make excellent pets
for the first-time pig owner.

should want to use the box. It's imperative that they don't lose their focus on what comes naturally by thinking of the box as a food dispenser. The box should be cleaned often and enlarged as the pig grows.

No other animals should be allowed to use the box. If the pig goes on the carpet, enzyme cleaners, such as Natures Miracle, will take the scent away. The enzymes actually "eat up" the urine. This is imperative to use or the pig will smell the old urine and think that this is the proper place to urinate again. Vinegar will not take away the scent well enough to fool the porker.

A soft child's blanket or two will make a nice pig bed. Pigs love to sleep. (Eating and sleeping are what they do best!) A sleeping bag can trap a pig, and some pigs have even been known to die from the distress of being trapped at the bottom and panicking. It is also important that your piglet be kept warm and safe from dogs.

When your piggy has rested a few hours, it is time to start socializing him. Going into his room for short sessions (10 minutes or less) often works well. Concentrate on teaching him his name, along with the command "come." Progress may be slow. Let the pig come to you; let it be his decision. Hold out your hand with popcorn, grapes, or pig feed in it. Say "treat" when you pay him. Repeating this will impart to the pig that he has been rewarded and has done well. It may take several days for the pig to come, but he will, eventually.

Try not to "crowd" the pig. It's important that he gains self-confidence and learns that he can trust humans. If you consistently reward him for approaching you, your piggy will be motivated to take chances in order to get to know you. Don't drop food on the floor. Make the pig actually take it from your hand. As he gains confidence, start scratching his back as he does this . . . pigs LOVE to be scratched! Saying "good pig!" when he comes will reinforce his behavior and set up a foundation for training.

Allowing the pig to always come to you, instead of your going to him, will make him realize that he does have some power and free will, and that humans can be beneficial to him. It can be a longer process if the breeder has done no socialization. You may want to hold the pig's food bowl in your hand as he eats. The pig will learn that you are responsible for feeding him and, subsequently,

he will learn to love you.

Being predictable and gentle will help, as well as putting him on a routine. Try feeding and socializing him at the same time every day, as well as putting him to bed and waking him up at the same time. Pigs don't like surprises when they are young. Having to chase the pig will set the training back because predators chase pigs. Instead, be slow and attempt to coax the pig.

Socializing Two (or More) Pigs

For optimum results, piglets, especially littermates, should be separated for socialization. This is because piglets tend to act as a herd rather than individuals if they are not alone. Instead of thinking, they tend to simply react and copy one another. One pig is usually dominant over the others, and the less dominant pigs simply do what the dominant pig does. They already have a leader; they don't need a human for the job. They won't listen to you very well as long as they have one another. They don't need you for companionship and security. You are just the one that hands out free food!

Certainly, groups of pigs can eventually be socialized and trained, but it takes longer. Separating them temporarily will open their awareness and make them open to responding to your training. After they have been socialized, they can be put back together.

Lifting a Pig

It is important when lifting a pig that his head and body are at the same level. Support his neck and rear end. Make him feel secure that a predator is not grabbing him. Do not hold him by his tummy; it's uncomfortable for him.

An effective way to impart to the pig that lifting is a GOOD thing is to pay him with treats when you pick him up at first. As with all pig training, going gradually and in increments is important, as is positive reinforcement. Pigs despise heights, so starting at the ground level will behoove you.

Getting down on the floor and gently lifting the pig to your lap works best. Be prepared for an ear-splitting scream. Cradle the pig evenly, and let him tuck his snout in the crook of your elbow for security. If you hold him at an even level, it seems more like the floor to him. If he is not too excited to eat, pay him.

Allow his feet to dangle while you support his body. If he becomes extremely distressed, put him down and try again in a few hours. Make the training sessions short.

As always, using positive reinforcement works well with porkers. You may wish to feed your pig his entire meal while he is in your arms. The pig will soon learn that he is fed when he's in your lap. As he becomes more secure, increase the height and give fewer treats. It is an excellent idea to say "up" when you lift the pig and "down" when you set him back down on the floor. Pigs hate surprises and are quite capable of understanding commands in English. This will prepare him so he won't be frightened.

Bonding

Bonding to your new pig can be accomplished by training. Training allows you to spend time with your pig

and to provide him with challenges. It stimulates his mind and makes him respect you as his leader, someone to trust and look up to. Training is the most humane way to accomplish trust.

Bonding can also be accomplished by spending a lot of time with your pig, giving belly rubs and ear scratches. Talk to your pig. It is amazing what he can understand. Try not to always be feeding your pig, unless he earns the food. In other words, NO FREE LUNCH. If he begs, ignore him. You should be in charge of when the pig is given food.

Always make training sessions a positive experience. End on a happy note by doing an easy trick that the pig has mastered well. Keep your pig's tail wagging and keep the sessions short. Pigs have very, very short attention spans. Some piglets can focus on one thing for only a few seconds, so piglets should be trained for only a few minutes at a time. As the pig gets older, sessions can be increased to 20 minutes or more.

Strangely, training a new piglet to perform tricks seems to take away his fear of humans. The pig focuses on the feat and the reward instead of his fears. He learns to look up at you, awaiting the next command. Pigs love learning tricks because the tricks are like games to the pig. Pigs are curious, mischievous animals that need a challenge and something to do, especially if they are housebound. Training the pig also seems to eliminate any destructive behavior that arises out of boredom.

CHAPTER 4

There's A Pig In My House!

After the new porcine arrival is no longer frightened, is willingly approaching his caretaker, is litterbox-trained, and comes when called, it is time to expand his world. The piggy should be ready to expand his small, confined area in about four weeks. Living in the house can be a challenge for a young pig. And it can be a nightmare for his owners if they desire to keep him in the house and not train him. It is up to you to solve any behavioral problems by introducing him gradually to the entire house. The pig has been taken from his natural environment, and you expect him to live by your rules. So you must patiently train the young piggy to be able to SURVIVE in your world.

Unfortunately, many miniature pigs are abandoned by their owners. This may be because the pig has grown larger than anticipated, or it may be because the pig was not adequately taught how to live in the home.

If the pig is not taught, on a daily basis, how to live in the abode, he will go by his own preferences. Pigs are creatures of habit, and once they start a particular habit, it can be hard to break later on. On the other hand, if the pig is imprinted with good habits right from the beginning, he has the potential to make an excellent pet.

Some pet owners keep their pigs outside. This will cut down on aggression and territorialism because the

pig has no territory that he must share with humans. There is no competition as to who owns the house. It is a personal preference whether to house the pig inside or outside. Some people desire a more "intimate" relationship with the pig, so they have purchased the pig as a housepet. However, a pig that lives inside needs more training and discipline. A house that is full of stress, chaos, and a lot of children and animals is not ideal for a porker to live in. Pigs like predictability and serenity. Small children are not always practical with a pig in a home either. Pigs can learn to be aggressive and demanding with children under eight years old if the children are not taught to stand up to the pig.

Pigs can teach us too. They are incredibly intelligent animals that are truly unique. They are neither dogs nor cats, and can't be trained or treated as such. Pigs live for the day, never looking ahead. If they had a choice, they would definitely eat dessert first! Pigs have a sense of humor and bring that out in people who love to make pig puns. Pigs are simple, unpretentious animals. They need shelter, food, and security. In addition, they should be exercised and stimulated. And most of all, pigs are PASSIONATE animals. They enjoy life to the fullest!

Teaching a pig to live in harmony with you in your home can be a challenging and enjoyable experience.

But first of all, the pig needs to be taught the physical skills to get around the house.

Pig Proofing the Home

Pigs are curious pets that love to knock things over and uproot objects just for the fun of it. They will also eat just about anything that can be ingested, including dog food, paper, wood chips, shoes, toxic houseplants, electric plugs, plastic bags, foil, etc. A new pig owner must anticipate any dangerous items that a piggy might be attracted to. Porkers will open cupboards, refrigerators, and overturn wastepaper baskets in their constant pursuit of the elusive, unknown food morsel.

So the piglet must be closely SUPERVISED until the house is totally pig-proofed, and he learns what is not acceptable to touch.

Stairs and Ramps

Going up and down stairs is a simple task for dogs and cats, but for pigs it is a physical challenge. Those short stubby legs, coupled with a pendulous belly, make a pig apprehensive about using stairs. Combine that with a pigmy's natural fear of heights and most people end up carrying their young piglets upstairs. Stair training should begin when the piglet is about three months old. Initially, he should be called and rewarded for climbing one stair. You may have to bait the pig at first, putting a morsel of coveted cheese on top of the step.

After he climbs the first step, say "good pig" and teach him to turn around and go back down. After he has mastered one step, increase the number of steps, and only dispense a treat at the very top or bottom. Once he learns this, instead of throwing the treat on the step, give it to him so that he is actually LEARNING a command instead of simply chasing treats. Slippery stairs may need to be carpeted in order for the pig to have proper traction.

Ramps should be used to get heavier pigs into cars or

up steep steps. The pig may not need a ramp now, but it will be much easier training a young pig to negotiate a ramp than waiting until he needs it to start training him. A smart pig will not go up a wobbly, slippery, steep, or narrow ramp. He is vulnerable to falling and injuring himself, so he will wisely refuse. A ramp should be at least 18 inches wide and should not wobble or shift once the pig's weight is on it. It should be carpeted or have wooden cross slats every 6 inches.

Once the sturdy ramp is built, place it on the floor, and bait your pig into walking on it, directing him to do it all the way until the end. Pay him at the very end with a special treat that he rarely gets and praise him lavishly.

For an entire week, have him walk on the ramp and pay him when he does. Then prop it up at a low angle (such as on the stairs) and retrain him. Increase the angle and only give him one treat after he steps off the ramp at the very end. If your pig seems reluctant to use the ramp to get in the car after this, help make him feel secure by leading him up on his harness. This is a good idea in any case because if your pig ever falls, it will set your training back immensely. Have a special surprise treat waiting in the car for your piggy.

Crate Training

It is wise to train a pig to be willing to go into a crate for vet visits and travel BEFORE he is actually required to do so. A large crate (kennel) can also be a good, private home for your pig to sleep in. Pigs that sleep in crates appear to act less territorial because they have their own little nest. A crate must always be large enough for

the pig to turn around in and assume his normal postures. Extremely large crates that are four feet long can be purchased.

Pigs are afraid to enter enclosed spaces. They could be trapped if a predator comes by and their only route to escape is blocked. Therefore, to introduce the porker to the crate, remove the top half and place a bowl of feed and familiar-smelling blankets inside it. Do not coax the pig into the crate. Ignore the whole situation and let the pig appraise the situation himself. The pig will enter the crate himself, even if it takes a few days.

Start feeding the pig in the crate. After a few days, put the top on the crate, put his bowl in it, and wait for the pig to enter. Never force a pig into a crate. Never! The pig will remember the experience and never want to re-enter the kennel again. Instead, put a treat in the crate and say, "Go into crate."

Never pull a pig out of a crate either. Use food and gently coax him out. Make the crate a good experience through the pig's eyes. Be patient. The pig will learn to love his crate if it's up to HIM when he uses it. If it's entirely his own choice whether to use the crate or not, the pig will see the kennel as a special, cozy refuge from the hectic world of humans. All you will have to do is ask him to enter or exit it when you want to use it for transportation.

Outdoor Potty Training

Potbellies prefer to eliminate outside instead of using messy litterboxes. Litterboxes can certainly be used for the life of your pig, but taking the pig outdoors is

Teaching "sit" will help you earn your pig's respect.

preferable. Pigs are naturally drawn to the sights and smells of the outdoors to eliminate. Training a young pig to go outdoors is relatively easy. Establishing a strict routine is desirable.

The young pig should immediately be taken outdoors first thing in the morning, after eating his meal, once or twice in the afternoon, the evening, and before going to bed. It is better to take him out too many times than too few. You are attempting to set a pattern that will evolve into a lifetime habit.

Rewarding the pig for a natural function (such as elimination) can lead to problems, both with litterbox training and outdoor training. Pigs love food, so if they are rewarded or coaxed with treats, they tend to get excited. The result may be that the pig goes outside expecting a treat and forgets to do his business. If he is rewarded for coming indoors afterwards, he may come in too soon in order to be rewarded faster. So keep the food at a bare minimum and use praise instead.

Taking the pig out every day at the same time will help insure success. As the pig gets older, you will be able to cut down the number of times he goes out.

Dog Doors

Pigs learn to use dog doors easily. If you buy one, be sure to buy a size that will fit your porker several years down the road. To train your pig to go outside through the door, go outside, put your hand through the door (holding the flap up), and call the pig. Pay him when he is all the way through, and then reverse the process for coming indoors. A very low dog door is desirable, for obvious reasons.

Harness and Leash Training

Potbellies are very different than dogs. They hate being restrained because it could be a predator doing the restraining. Therefore, harness training must be done in baby steps in order not to distress the pig. Training him in a small, carpeted room that he is familiar with is a good idea. An adjustable, potbelly A-line or figure-eight harness is appropriate, with a loop to go over the head and a loop to go behind the front legs. These harnesses are specially designed for potbellies and have flexible nylon straps. They can be purchased on the Internet. A dog collar or harness will not fit properly or be comfortable.

STEP 1 — The first step is to measure the pig's neck and girth (just behind and in front of the potbelly's front legs) and adjust the harness to fit before you attempt to put it on. Add a few inches for comfort, especially the first time. Place the harness on the floor, sprinkling pig feed over it. The pig will sniff the harness, associate it with pleasurable eating, and think of it as his friend. After doing this for several meals, place the harness on the floor, spread it out, and put food inside the harness loop circle that goes over his head. The pig will place his snout in the middle of the loop on the floor in order to eat the morsels.

STEP 2 — This puts him in an excellent position for you to slowly pull the loop over his head as he is eating away. His head is already in the center of the loop. Be slow and gentle as you ease it over the snout, the jowls, and finally around the pig's neck. Make sure that he still has food to focus on.

STEP 3 — Next, snap the back loop together right behind his front legs. The pig may be upset, so have special treats like grapes or cheese in order to distract him. You want to DESENSITIZE him to the pressure of the harness. At first, he will feel like he is being pulled or restrained by the pressure of the harness around his body. Talking to him in a soothing voice and reassuring him will help. Do not make the harness very tight the first few times.

STEP 4 — Attach the leash and follow your pig wherever he goes. Allow the pig to lead you because at this stage you do not want to pull on the leash and upset the porker. If the piggy gets too wild, unsnap the leash after a few minutes and take up the training again in a few hours. Otherwise, ask the pig to come to you as you walk along, holding the slack leash. Reward him and do not apply constant pressure on the leash. Instead, use light jerk-release pressure, so the pig does not feel totally restrained. If the pig seems upset, you are going too fast. If the pig backs up, go with him. This is not supposed to be a battle of wills. It is supposed to be a pleasant, educational experience. Until he is desensitized to the pressure of the harness and leash, letting the pig have his way will cut down on his negativism. If the pig runs, run with him. Otherwise, the pig will be suddenly jarred when he runs out of leash.

STEP 5 — Practice the leash and harness training daily. Start to teach the pig to follow you by saying, "Follow." Proceed only a few steps, say "good pig" and pay him while saying "treat." Gradually increase the amount of feet that he will follow you. Then start giving the leash

a short tug when you issue the command "follow." The pig will learn that he is to follow you whenever the leash is pulled.

All pigs should be leash and harness trained, not only for their own safety but because it teaches the pig that you are his leader — a person to respect and obey. And it allows you to take the pig places, expanding his world and educating the public about these exotic critters. No pig should be harness and/or leash trained if the weather is hot or even very warm.

If the pig seems extremely uncomfortable, remove the harness and end the session. Some flops and spins are normal, but if the pig is totally out-of-control, cease the training until later. A leash should never be used as a substitute to supervising a pig. Pigs cannot be staked out on a rope or leash and will panic if left alone on a tether.

CHAPTER 5

PIG PROGRAMMING, SIMPLE AND FUN!

Training pigs is certainly very different from working with dogs because they each have unique personalities and physical traits. Dogs are very energetic for most of their lives compared to porkers.

Pigs start out that way, but after age one, they are often slow moving and difficult to motivate compared to the energetic, agile dogs. Potbellied pigs need a special training system that takes into account their personalities and their uniqueness.

I suspect that pigs may have slow metabolisms because they were selectively bred to be fat and grow quickly in order to contribute to the food chain in Vietnam. And potbellied pigs are not as athletic as dogs. They are "mounted" on short stubby legs with a sway back and a large pendulous belly . . . a chiropractor's nightmare!

A potbelly may have some physical limitations compared to that of a dog. However, pigs can still be taught such athletic behaviors as jumping through hoops or climbing teeter-totters if they are highly motivated and not obese. They are even capable of taking on a dog's agility course in some cases.

Unlike dogs, pigs do not necessarily perform tricks to

please their owners at first. Pigs do it for the ultimate reward, which is usually food, and for the fun of it, generally speaking. (There may be exceptions to this.) They are highly motivated by food goodies, which is no surprise to anyone.

They want to expend the least energy possible in order to get the treat at the end. They are certainly not above cheating and actually appear to enjoy playing a "game" to see if they can fool the unsuspecting trainer into dispensing a treat for a job only half done.

An example of this occurs when training a pig to jump through a hoop. After the pig has proficiently mastered the task, it is not unusual for him to attempt to go UNDER the hoop, hoping that he will still be paid! If that doesn't work, the pig will attempt to go around the hoop or knock it over.

So the trainer can never let his guard down or lower his standards. Once a pig has mastered a trick or obedience lesson, it does not mean that he will do it forever in the correct manner.

Pigs have a very short focus or attention span compared to canines. They get bored very easily, especially if there are a lot of repetitions. And pigs do not like to fail. So it's a challenge for the trainer to keep the sessions moving quickly, yet not go beyond the pig's capabilities.

Pigs love challenges, games, and (of course) eating. So even though they have short attention spans and are not natural athletes, they are ideal candidates to be trained. This can be accomplished through using a simple and direct method, Pig Programming.

The Method: Pig Programming

Ten years ago, when we got our first potbelly pig, there was very little information about these newly popular, porcine imports. It was difficult to find a book about how to feed them, let alone train them.

We tried various methods using bells, clickers, targets, etc. Dog-training manuals or other animal-training books just didn't apply to shaping pigs' behavior.

I began to notice that the best results seemed to occur when I simply spoke to the pigs in good, old-fashioned English just as you would to a child. A few hand signals mixed in and the pigs were able to understand what I was trying to tell them very well. The pigs were capable of remembering the commands and carrying out the desired behavior. They enjoyed being trained almost as much as eating supper.

We developed our own piggy-tailored training method and called it "PIG PROGRAMMING." It was designed to be easily learned and enjoyable. Although we train our pigs to perform professionally, as a stage act, this method works equally well for pet pigs' training and obedience.

Pig Programming has three parts:

1. The Command
2. The Reinforcer Cue
3. The Payment Cue

Your Voice Says It All

In order to train an animal, we must communicate. An animal cannot read your mind, even the sensitive

This pig is into mischief!

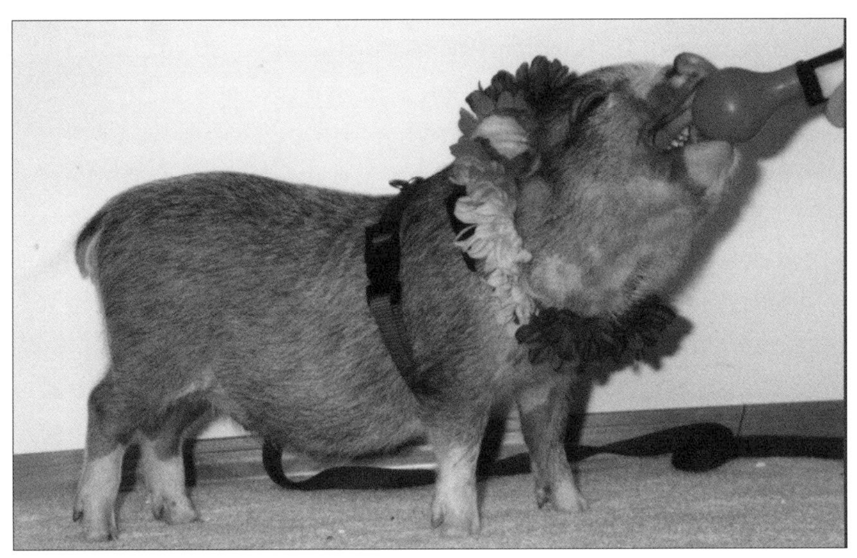

The Duchess of Pork is being trained using
Pig Programming (operant conditioning).

potbellied pig. Your voice is probably the most important training tool. For training pigs, it should be your major communication, telling the pig what you want him to do by specific words, tone, and inflection. Pigs will respond accordingly when your voice is used CORRECTLY. Your voice should impart to the pigs what you expect of them. It should also motivate them. A calm, soothing voice will elicit a different response from your pig than a high-pitched, silly tone, for example. A deep, authoritative voice will evoke a different reaction too.

Use a firm, low voice to get your pig's attention and respect.

Use an upbeat, appreciative, higher-sounding voice to reward or praise him.

Use an authoritative, deep voice to issue a command for training.

Use a low, gutteral voice to correct your pig.

Use an enthusiastic voice to motivate.

It is very important not to sound like you are asking your porker a question when you are giving a command. If you act like you don't expect the pig to obey (by the sound of your voice), your pig will not take you seriously. Using a low, firm-sounding tone will tell the pig that you expect him to obey. Do not let your voice trail off or die down. Try to sound like your grade-school principal when giving commands!

Women have higher voices than men and may sound less authoritative. Therefore, a woman should lower her voice and pay special attention to coming across as very intense and confident. Try to impart to the pig that you mean business by your voice inflections.

The Three Elements of Pig Programming

NUMBER 1: The All-Important Command

The command will let your pig know what specific action is required of him. Using very short, simple, direct phrases or verbs works well with pigs. For example, "sit," "jump," "go in the crate," "fetch the ball," etc. You will confuse the pig and dilute the impact of the command if you string out useless words around the command or ask the pig a question, such as "Porky, do you want to run and fetch the ball now?" or "Would Porky like to sit pretty?"

Remember, you are not asking your pig if he wants to do the task, you are telling him. Otherwise, your success will be very limited if the pig has a choice in the matter. It's good to have high expectations of the pig and show it in the command.

Make sure that the porker can hear the command, and do not repeat it over and over. Use it sparingly; attach power and importance to it. It is prudent to use the same word for the same command, always. Varying it will only confuse the piggy.

NUMBER 2: The Reinforcer Cue

After your pig has followed your command, he needs to know exactly when he has completed the desired action successfully. This cue will act as a way to communicate two things to the pig. It will tell the pig that (1) he can now stop whatever he is doing and (2) that he has been successful in doing the behavior correctly. An extremely short phrase, such as "good pig" or "good, Porky," is ideal. These phrases must always be the same. By using it, it allows the pig to relax and prepare for the next command or his reward if he is going to receive one. It releases him from the obligation of the first command and tells him that he has performed well.

When initially training a pig, the reinforcer cue should be given at virtually the EXACT SECOND that the pig has accomplished what he was supposed to do. This is because when a pig is first learning a new feat, he needs a cue that will let him know he has mastered the specific action you are trying to teach him. He will remember it and be able to do it again upon command.

If you want to impart to the pig that he is doing what you have asked him to do, but you want him to continue doing it, repeat "good pig" over and over until it's time to pay him.

NUMBER 3: The Payment Cue

The verbal payment cue will communicate to the pig that he has been rewarded for being successful and that it is time to move on. He will not be rewarded again for that particular completed feat after the cue is given.

There must be some verbal communication in order for the pig to realize that he has been paid all he is going to get for that feat and that he has completed the command successfully. Otherwise, the greedy, clever pig will simply stand there waiting to be paid again. The payment cue prepares the potbelly to listen

carefully for the next command and ensures that he will return his complete attention to you, the trainer. It acts as a separator and release between commands.

An excellent payment cue to give is "treat" or say "that's it." Say it AS you are giving Porky a reward.

Clicker Training and Pig Programming

Clicker training is totally compatible with Pig Programming. The clicker noise can be substituted for the reinforcer cue. Instead of saying "good pig," simply press a clicker mechanism. The pig will soon learn that the clicker noise means that he has done well, and he will be paid right after he hears it.

Clicker training has been around for decades. It was, and still is, sometimes used by trainers when their animals are in front of cameras. Instead of shouting at the animals, which could be a chaotic situation if there were several trainers, clicker noisemakers were passed out on the set to impart to the animal that he had done whatever was required of him. In the movie *Babe*, the animals were trained to respond to different sounds. These included clickers, pennywhistles, buzzers, and their masters' voices.

However, saying "good pig" instead will work equally well, if not better. Because pigs can be motivated by the enthusiasm in your voice, using your own voice, instead of a clicker noise, will give you a training advantage.

CHAPTER 6

THE PIG PROGRAMMING PHILOSOPHY

The Pig Programming method of training is designed to make teaching your pig simple and fun. In the end, the potbelly will be the benefactor. If your pig has destructive habits, such as chewing furniture, rooting up carpets, dumping over wastebaskets, or is just plain bored, training him will take his focus away from mischief and allow him to expand his horizons by overcoming challenges in a positive way.

Pig Programming's intentions are to help the pig learn quickly and easily. It is intended to make the pig think that the trainer and he are playing a game, although it is a serious game.

When training any animal, it always behooves the trainer to concentrate on the natural tendencies that the specific animal has and to use them to his or her advantage. For example, dogs are natural retrievers, so a canine trainer would use this ability (fetching) to teach the dog several feats. Pigs are natural rooters and pushers. This makes the feats that require pushing or nosing ideal choices to use as tricks to teach them. Playing the piano with a pig's snout is a good example of this. It is always best to go with the natural talents of the animal when first starting to train.

The advanced feats are often tricks that go against a pig's natural talents, such as athletic endeavors, technical feats, jumping, or feats that require carrying objects (pigs are not natural retrievers).

Pig Programming utilizes the age-old training concept of CONDITIONED RESPONSE. Remember Pavlov's dogs who "learned" to salivate when a bell rang? When a pig is consistently encouraged and rewarded to perform a specific behavior, over a period of time, he will ultimately instigate that behavior on his own if trained properly.

Pig Programming believes only in POSITIVE REINFORCEMENT. A pig is never punished for not performing a particular behavior. He may have a time out, but this is not meant to punish the pig. It is simply a rest and relaxation period for both trainer and animal. Pigs do not respond well to negativism or intimidation. They tend to rebel, pout, and/or get frightened, or act neurotic. It sets the training back if the pig is not enjoying himself. An angry or upset pig does not learn. Therefore, it is up to YOU to motivate the pig.

We've already talked about pigs' short attention spans. Some young pigs have attention spans of only a few minutes. Because of this, Pig Programming divides each feat into virtually the smallest increment possible. This way, each segment is easier to focus on because it involves less time. The pig gradually learns each step separately but in the correct order, and then these pieces are put together to form the feat in its entirety.

Pigs are not the easiest animals to motivate. They can be sluggish with selective hearing. Thank goodness for food! But your attitude toward training is important too. If you are enthusiastic, your pig will be enthusiastic. If you act

like a leader, your potbelly will act like a follower. Your posture, your voice, and your demeanor play a huge role in whether the pig is motivated to succeed or is just passing time.

Observing your pet and being in tune with him are important too. Is your pig frustrated? Should he take a time out? Is the pig not feeling up to par today? Are you going too fast? Should you back up?

Training can be frustrating and time-consuming, but pigs are incredibly intelligent animals that need to use their God-given intellect. The more a pig learns, the more he WANTS to learn. Training almost literally brings a pig to life.

Have fun with it!

CHAPTER 7
OBEDIENCE AND TRAINING BASICS

Verbal cues will give the potbellied pig structure and allow you to communicate with him without confusion. With both obedience and tricks, it is important that the pig be issued a succinct command. Then the pig should be quickly and verbally reinforced when he does what is asked of him ("good pig!"). Finally, the pig should be released and paid at the same time the trainer says "treat."

The Reward

Small, easy-to-handle treats, such as popcorn or dog food, are ideal. In addition, your pig should get verbal praise and a pat on the head. As a general rule, do not drop the treats on the floor. Instead, hand them to the pig. This gives you more control, and lets the potbelly know that he has earned the treats and that they did not just fall from the sky onto the floor. One kernel of popcorn or one piece of dog food per trick will suffice. If your pig is overly portly, air-popped popcorn or geriatric, low-fat dog foods are appropriate.

The Training Room

The potbelly should be trained in a large, quiet room where there are no distractions, such as children, TV sets, other pets, or other people.

Proper Timing

Many trainers of pigs make the mistake of training the porker when his tummy is empty, just before meals. This is not beneficial because the pig will not concentrate on you or take direction. His internal time clock has told him that he has been "gypped" out of a meal, and he will be in no mood to focus on what you are attempting to teach him.

Instead, train the pig several hours after he has been fed. Pigs should only be trained for a few minutes at first. Gradually, increase the time to 20 minutes per session.

Obedience

Teaching obedience will impart to the potbelly that you are "Top Hog" in his herd. Once he has been taught obedience, you may notice a change in your porker's behavior. He will seem to respect you more in many cases. Training obedience is the most humane way to impart to your pig that you are at the head of the family social order.

This piglet is well adjusted. Getting your pig used to your home should be done gradually.

Use food to make your pig learn to enjoy being held.

Teaching "Sit"

Teaching your porker to sit is an easy feat, even for piglets. Place your pig with his back close to a wall or corner.

Say "sit" (command) and take a treat in your hand, holding it just out of the pig's reach above his head.

As the pig raises his snout to smell the treat, move the treat toward the back of the pig's head, stopping just above his eyes. As you move the treat back, just out of reach, it should force the pig to sit. The second that his rump is level on the ground, say "good pig!" (reinforcer cue) and pay him, saying "treat!" (payment cue). Practicing next to a wall will prevent the porker from backing up without squatting.

Teaching "Stay"

Because of a pig's short attention span and food lust, teaching a pig to stay is much more difficult than teaching a dog to stay. The greedy pig wants to jump up for his reward, thus breaking the stay.

Using hand signals are very effective. Pigs are very sensitive to motion, so hand signals are a plus with any trick. It doesn't matter what kind of hand signals you use as long as you are consistent.

Put the pig in a "sit" position. Give him the command "stay." Keep repeating the word "stay" for a few seconds and have your arm extended, palm out, pumping your arm toward the pig as if you are pushing him back. After only a few seconds, pay him and say "treat!"

Very gradually, increase the amount of time used to make the pumping motion. If the pig breaks the stay,

start all over again. Start by standing 2 feet away from the pig during the stay, and gradually increase the distance between yourself and the porkster.

The secret to teaching the pig the long-distance stay is to walk over to the pig to pay him instead of asking the pig to come to you. If the pig realizes that he won't be paid if he comes to you, he will have no reason to break the stay. He will wait patiently until he hears the word "treat" and you have approached him.

Teaching "Lie" or "Play Dead"

Pigs do not easily lie down on command at first. The reason is that pigs are preyed upon and it puts them in a vulnerable position. Pigs like to be standing on their feet in case there is trouble, and they need to run and hide. They will lie, but it has to be the pig's choice. We have found that the best way to teach a pig to lie down is to give him a few scratches on his belly. He will flop over, and as he does, say "lie." Then feed him as he is in the prone position for a few minutes and say "good pig!" several times. Pay him again, saying "treat!" while you do.

This is a slight modification of the Pig Programming method because the pig is paid both as he is doing the trick and afterwards too. But soon the pig will learn to lie down without being rewarded at ground level and should only be paid at the end.

Teaching "Bow" or "Curtsy"

Kneeling is an excellent feat for a pig to learn because it comes to him naturally. When piglets nurse on the

sow, they fold their front legs under them while standing on their rear legs. Follow the Pig Programming verbal steps and ask your pig to "come." Place some pellets of pig food on the ground, and as the pig chomps away, fold the pig's front legs under him, using both of your hands. Quickly release his legs and pay him with a special treat. Practice with the potbelly, folding his legs under him as you say "bow," several times a day.

A pig can also be taught to curtsy by a variation of the same method. Simply fold one leg under, instead of two. In no time at all, the porker will be able to bend his knee in respect to royalty's presence!

CHAPTER 8

TEACHING BEGINNING TRICKS

Teaching your pig tricks will not only often solve porkers' problems with boredom, aggression, and lack of obedience, but it will also allow you to show off your pig and educate the public. There have been many myths and misconceptions about potbellied pigs, and if you get a chance to display your "ham" doing amazing feats, you are also educating the public about what these bristly critters are all about.

Nursing homes, schools, and libraries would be tickled to have a visit from your talented piggy!

Always use the Pig Programming verbal cues: the command, the reinforcer cue, and the payment cue. Props for many of the tricks can be purchased at large toy stores. Strolling through toy displays and browsing will give you new ideas for training as you look at the various contraptions for two- to five-year-olds. These toys are made of sturdy plastic and can withstand hoof scrapes and overly enthusiastic nose pushes.

Be creative!

The beginning tricks will teach your pig the basic moves he needs to know to accomplish the more technical feats. He will have a foundation to draw from and should recognize basic commands.

Most difficult tricks are based on the beginner concepts of push, fetch, pull, up, and down.

The Spin and Figure Eight

Most pigs learn the spin by themselves! To train your pig, simply bait him with food (holding a treat above his head) to rotate his body in a circle. He will look up and follow your hand's circular motion, causing him to pirouette. After he gets the idea, do not bait him again. Just have him follow your empty hand in a circular motion and only pay him at the end. Teach him the opposite direction too.

Ideally, he should eventually be able to do it without any use of your hands to get him going in a circular direction. It is more impressive if you teach the porker to automatically spin by himself when you give the command, rather than starting him out by using your hand to lead him. This will take practice.

The figure eight can be taught by spreading your legs at least 3 feet apart while standing. The idea is to have the pig weave through your legs in a pattern. Bait the pig to move through them (by holding a treat above his head) in a figure-eight pattern. Verbally reinforce him, and teach him to go around several times. After he's learned, stop the baiting and only pay him once at the very end. Then increase the times you ask him to weave but still pay him only once. Otherwise he will stop the flow, expecting a goody.

Let's Get Musical

Playing a bicycle horn is a natural for a pigster because it involves using his mouth. Place part of the noisemaker, rubber bulb in the pig's mouth and say "play."

Tuff Hank giving the "high four."

It takes pigs, like Nellie, several weeks, at least, to learn to fetch.

The very second that the horn makes a slight noise, say "good pig" and pay the pig. Even if no noise is made, still pay the pig. He will get the idea that touching the bulb with his mouth is profitable. Repeat this action over and over. The pig will get the idea that he is supposed to make noise. After a few times, he will be honking up a storm!

Playing a child's low-height piano is simple for a pig. Place a few kernels of popcorn on the keys. The very second that the pig makes a piano noise by eating the popcorn, say "good pig" and pay him. Keep practicing and eventually remove the popcorn from the keys. The pig will give you a Chopsticks recital on his own.

The Shake and Wave

Gently pick up your pig's front leg and pay him. At first, he will think you are crazy. Once he is comfortable with your lifting his leg, repeat the action and pump the leg up and down. Say "shake," paying him afterwards.

To teach him to wave, simply practice with him a long time on shaking hands . . . or legs in the pig's case.

Eventually, he will raise his front leg all by himself. When he does this, make a hand signal like you are waving back to him while saying "wave." After a while (and it may take some time) he will be doing this all by himself, at odd moments, trying to get paid!

Roll Out the Red Carpet!

Teaching a pig to roll out a long, narrow carpet is easy because it involves the pigger's basic tool, his snout.

Put popcorn kernels, spaced evenly, inside a rolled-up carpet. Show the pig the first kernel, and he will roll the rug out in nothing flat because he smells the rest of the popcorn. Pay him at the end. Practice again and again, eventually eliminating using the popcorn inside the carpet. Only pay him at the end, after the entire carpet has been unrolled.

Fetch

It can be harder to train a pig to fetch than it is to train a dog. The pig thinks that it's stupid to carry or hold a bone in his mouth. He'd rather reserve his mouth for eating! Pig Programming comes into effect here as it does with most complicated tricks having more than one concept to teach. We will break the trick down into small segments, as Pig Programming dictates, in order to better convey to the pig what we want him to do.

Put a 6-inch long, plastic dog bone in the pig's mouth. As it's actually in his mouth, say "good pig." Wait until the pig is comfortable with your holding the bone in his mouth, then continue on to the next step. Place the bone on the ground, right next to the pig, and show it to him. The pig should pick it up. (If he doesn't, back up to the previous step.) When he does pick it up, immediately say "good pig" and grab it out of his mouth. If you don't grab it, he will learn to drop it at this early stage. Increase the amount of time that the bone is in the pig's mouth before you take it away from him.

Also, gradually increase the distance, inch by inch, that the bone is away from the porkster. He will learn to bring it to you and drop it in your hand in order to be rewarded. However, a pig's naturally blurry eyesight will probably limit the distance that he can fetch to less than 20 feet in most cases.

A piggy can learn to fetch many objects. He can bring in the newspaper, fetch a Frisbee, carry a basket, or even place your slippers next to your easy chair! There is no limit to the number of tricks that a pig can do, once he learns to fetch.

Mowing the Lawn

Pigs are natural landscapers! Buy an inexpensive, plastic, play lawnmower. Put a piece of popcorn on the back of the mower at the pig's eye level. Tell the pig to "push lawnmower." As he tries to get the popcorn, the mower should move an inch or two. The second that the mower moves, enthusiastically say "good pig" and pay him. Soon he will figure out that moving the mower is profitable! Mowing the lawn will teach the pig to push and many tricks are based on that principle.

CHAPTER 9

INTERMEDIATE AND ADVANCED TRICKS

Golf

In order to teach your potbelly to be proficient at golf, it is prudent to use Pig Programming and break the trick down into segments. First, teach Porky to push a ball by placing a piece of popcorn under a soft ball and saying "push." The instant he moves the ball with his snout in order to find the treat, say "good pig" and pay him. Repeat the procedure, until he learns to move the ball across the entire room, with a treat delivered only at the end of the feat.

Using a golfer's putting ring, place the ball right in front of it. As the ball is moved toward the ring by the porkster's snout, reinforce the pig. Eventually, the pig will get the ball in the cup, probably by chance at first. Immediately praise him! Show enthusiasm!

This trick will take some practice, but eventually, the pig will realize that the ball must go into the cup before he is paid.

Soccer and Bowling

The pig has already mastered moving balls and the concept of pushing objects. This time, we will place the

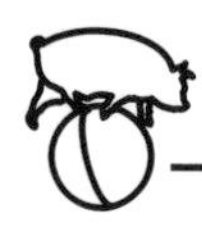

bowling pins or soccer net in front of the pig with the ball in between the pig and the ultimate goal. Make the distance only about one foot that the pig has to push the ball. Reinforce the pig for pushing the ball toward the net or pins, and pay him through the net or over the top of the pins. This is crucial because the pig should identify being paid with the net or the pins, and not you. (You want him to move toward these goals with the ball and not move toward your body instead.) Once he has this down, gradually increase the distance of the ball from the net or pins. Your pig will become a sports fanatic!

Shaking Head "Yes" or "No"

To teach the pigger to make an up and down head motion, signifying "yes," simply take a treat and bait him to look up and then down. Reinforce and pay him and then practice. He will get the idea that looking up and down and moving his head is what you are asking of him.

To get the porker to shake his head, signifying "no," gently touch his ear. The pig will shake his head, and you should immediately use the reinforcer cue and pay him. This will take some practice.

Playing Six Horns

Once Porky has learned to sound a single horn, mount six horns on a low stand. Command him to play the first horn and then pay him. Very gradually, increase the number of horns, always in the proper order, from right to left. When he plays them *all*, in order, pay him when the last horn has sounded.

Pigs will sing for their supper as Truffles demonstrates.

The Vietnamese Crawl

Put the potbelly in the bow position. Get down on the floor, very low, with a treat, holding it in front of his face. Move it out of his reach. He will crawl a step or two, attempting to eat it. Immediately, use the reinforcer cue and pay him. Gradually, increase the crawling distance.

The Slam Dunk

Teach your pig to fetch a pig-sized basketball the same way that you taught him to fetch a plastic bone. Then, when the pig attempts to bring the basketball to you, have a hoop handy and place it directly in front of you. Knock the ball out of the pig's mouth, gently, into the hoop and immediately reinforce him. Repeat, and the pig will get the idea that the ball belongs in the hoop.

The Piggy Bank

Purchase a large piggy bank and cut a hole in the top that will accommodate large plastic coins. Use the same principle as for a basketball, knocking the coin out of the pig's mouth when he is standing over the top of the bank. In no time at all, he will be saving for his college education!

Hoop Jumping

Purchase a hula-hoop and bait your pig into walking through it as you hold the bottom of it on the ground. Gradually, raise the hoop, reinforcing and paying the pig for going over it. If he refuses, back up. Soon, he should be proving that pigs really DO fly!

"This Little Piggy Went to Market"

Purchase a plastic toy grocery cart. If your pig is agile, bait him into standing on his hind legs with his front feet on the cart handle. After he is comfortable doing this, get in front of him and put a treat under his nose as he is standing at the cart. Retreat one step, so the treat is just out of the pigster's reach. He should move forward and, by doing so, move the cart too. Immediately issue the reinforcer cue and pay him. Gradually, increase the distance that the pig is expected to move the cart.

If your pig is not athletic, he can push the cart while standing on all fours. Touch the back of the cart to your pig's snout, give him the reinforcer cue, and pay him. He will want to touch the cart and will eventually end up moving it a few inches forward. When he does, be very enthusiastic in your praise and pay him. Soon, he will be doing his own grocery shopping!

As your potbelly learns more and more feats, he will get into the rhythm of Pig Programming. He will wait for a short command, listen for the reinforcer cue (e.g. "good pig"), and then wait to hear the payment cue. Never tell the pig after he has done a trick "good sit," "good fetch," "good spin," etc. Doing so will detract from the training program. The command "sit," for example, should only be vocalized if you actually want the pig to sit. Saying "good sit" instead of "good pig" dilutes the original command. A pig is intelligent enough to realize that if you ask him to sit and then you immediately say "good pig," you are referring to the sit.

CHAPTER 10

AGGRESSIVE BEHAVIOR IN HOUSEPIGS

Potbellied pigs generally have very sweet, predictable, laid-back personalities compared to other animals. They are gentle, warm creatures by nature and demand very little except a soft bed, plenty of food, the chance to forage, and the opportunity to sleep most of the day. They can be very affectionate toward humans, and owners bond to them easily and very deeply. They are truly remarkable, unique, loveable animals, but sometimes misunderstood because they are such "new" additions to the list of companion pets.

Like all animals, they can show some hostility if frightened, threatened, or confused. Most potbellied pigs show virtually no aggressive tendencies. Potbellied pigs have the potential to ACT aggressively at times, but they are truly not aggressive animals as such. They are not predators; they are the ones preyed upon.

Since they are herd animals, it may be a threat to them if their "herd" and their position in the family group are not safe and well established. Their herd has a pecking order, and the pig should not be confused about what his place in that order is, or he may be unfriendly toward houseguests and the befuddled owner too. When a potbelly comes to live in a home, the family becomes his new herd. There will be newly established rankings,

and the pig will learn from the family members just what slot he occupies. If there is confusion over who belongs where, he may start to physically challenge household members to prove that they deserve their status. In other words, the pig will attempt to move up the social ladder. Submissive members, like children or houseguests, will be challenged first.

This competition can cause inappropriate behavior in companion pigs, which can be very upsetting and challenging for the owners to correct. For the "only" potbellied pig that lives inside the home, this is especially so. Pigs that live outside rarely have problems with this type of behavior, nor do pigs that live with other pigs in the house. Your "only" pig may be docile with you, but may occasionally attempt to nip or charge houseguests. Or he may act obnoxiously toward you, the owner, AND occasionally the people who visit your home. Fortunately, a pig is neither a good biter nor a particularly strong opponent. But still, it is disturbing to owners who adore their "baby" to see him attempt to attack someone for no apparent reason. And, often, traditional discipline fails to correct these mysterious outbursts.

It is important to seek professional help if your pig is blatantly hostile and nipping. There are some "remedies" for aggressive behavior that can make your animal actually worse. Punishing the pigs physically seems to breed hostility in mini-pigs. If the owner acts extremely punitive toward the pig, it can be like adding fuel to the proverbial fire. Pigs should never be subjected to painful punishment.

Instead, the cause of the problem must be addressed. Once the cause of the problem is taken care of, the hostility will go away rather quickly.

Top Hog

We already know that pigs are herd animals. There is a pecking order from top to bottom with the strongest porker being "Top Hog." The herd order is maintained by body language. Slots are sometimes earned by ardent fighting, posturing, and charging. Once a ranking has been earned, a pig must defend it forever. If he appears vulnerable to a pig in a lower slot, he will be challenged. If he gets weak or ill, he will gravitate to the bottom of the social order. So it's important to be CONSISTENT in acting like the head of the herd.

When your piglet is initially brought home, he will form an opinion of you based on your strength as a leader. He will look at you either as a littermate or as a mother. If he sees you as a littermate, you are someone to play and spar with. A littermate is someone to shove around and compete with. He may nip at you, he may swing his head at you, and he may charge you. That's what littermates do.

If he sees you as a mother, he will respect you, listen to you, and be in awe of you. He will give you the power to set the house rules. He will not challenge you because he has accepted his role as the "underpig" or omega animal. He will be content with his pecking order slot and willing to learn and take direction. He will be a gentle, tractable animal as most potbellies are.

A demanding piglet will grow into a hostile-acting adult. Piglets that are weaned too early (under five weeks) are also prone to aggressive behavior. They don't learn to act like pigs because they missed out on the sow's disciplining. And they are easy to spoil because they are

Pigs can be overly demanding around children.

so diminutive. It is very important that all very challenging actions be nipped in the bud, when the piglet is young, so that he does not "bite the hand that feeds him." If a person can learn to "think" like a pig and take the proper steps, the potbellied pig has the potential to be a wonderful roommate.

The Spoiled Housepig

Miniature pigs are delightful animals. They can be affectionate, they are intelligent, and they can be playful and sensitive. It is difficult for some of us not to fall completely in love with them, but sometimes love is not enough. They have almost human-like personalities and act like inquisitive two-year-olds. When a piglet is first brought into the house for socialization, he is fearful and humble. After he is socialized, it is very easy to start giving into the demands of the new companion pet. It is hard to imagine a sweet, 10-pound animal being aggressive and competing with us for territory. It can be difficult, at first, to recognize that the pig is training the owner, not the reverse. And subtle signs of impending undesirable behavior are often ignored or excuses are made.

A pig is an observant, clever animal that is able to read signals. When he screams for his meal in the morning, begs for food and gets it, or refuses to go outside to potty and gets away with it, he is getting signals that he is Top Hog. When a child acts afraid of a pig or gives in to him, it sends a signal to the pig that he should move up in the pecking order. When he swings his head at his owner and the owner retreats, or he is allowed the full

run of the house with no rules or boundaries, the porker will gradually come to realize that he has power over humans.

There are many subtle ways that the pig learns that he has an opportunity to take control of the family hierarchy. When interacting with a pig, the owner must ask himself, "How will this action impact the pig in the FUTURE?"

Pigs are easy to spoil — especially housepigs. Many owners enjoy over-feeding them and giving in to the pig's whims. Spoiling often leads to demanding behavior. And once a pig starts demanding, the sky's the limit. The porker has started to take control. It is the owner's job to impart to the pig that he does not NEED control. His owner will take control, be the pig's leader, and make the pig secure and happy in his environment. Once this has occurred, the pig will be happy and content to have someone else take care of HIM. Once the pig is content in his slot, the pressure is off, and he is free to be the docile animal that all potbellies have the potential to be. He will no longer be frustrated. It's the frustration and confusion that originally led to the hostile behavior toward houseguests or family members.

Fair Discipline

Disciplining a pig is different than disciplining a dog. Pigs are animals that are preyed upon, whereas dogs are predators. If you frighten your angered pig by hurting him, he will neither forget nor forgive you. And he won't like you either, perhaps forever. The pig must always be cautious that he is in a safe environment, where no one will hurt him. He does not have the physical power to fight back like dogs do.

Pigs demand fair discipline. They have no inborn sense of right and wrong. We must teach them. A pig that is pestering a houseguest will not understand why he is being punished. He is behaving naturally and taking control of his herd. So instead of punishment, addressing the original cause of the behavior will produce the best results.

An angry or adrenaline-fueled pig will not listen to reason anyway if overly strong punishment frightens him. Shouting at the frustrated porker, squirting water, or hitting the pig are relatively useless tactics in most cases. This presents a challenging situation to the concerned owner. Threatening behavior can't be tolerated, but what can be done to stop it, at the very moment that it occurs, and teach the pig not to repeat it?

Any blatantly aggressive-acting animal must be removed from the room, immediately, after telling him "NO." A pig that threatens to nip or charge is being powered by adrenaline, a chemical that gives him a rush. The pig must be sequestered until the rush goes away and he does not endanger people or himself. Ideally, this should be at least a half-hour timeout. This timeout will not teach the pig to not be hostile. This must be done to impart to the pig, on a DAILY basis, that you are "Top Hog" and the pig's leader.

The entire family must implement a plan in order to take away the confusion in the social order, once and for all.

CHAPTER 11

The Ten-Step Program to Discouraging Aggressive Behavior

Most owners who have watched their pigs bullying guests or snapping at them have exhausted the traditional methods of discipline. Not only are pigs unique in the way they look, but they also have unique personalities. In the past, some owners have tried screaming at the pig, stomping their feet, acting like an enraged pig, grabbing snouts, or throwing water. One disgruntled caretaker even bit her own porker! Such tactics are usually not successful long-term because they treat only the symptom of a deeper cause: the pig's desire to be "Top Hog" and his confusion over where he belongs in the family.

The philosophy of the Ten-Step Program is not to change the pig, it is to change the owners. Subsequently, the pig will react differently TO the owners. The psychology of the caretakers must change. It would be very difficult to alter the deep-rooted herd instinct of the pig. The pig is behaving naturally and doing what pigs are supposed to do. He is taking over the responsibility of leading the family herd because the family members have defaulted by spoiling him. No one else is acting like a leader, so he is reluctantly taking over the job.

The Ten-Step Program is the humane and effective way to send a message to the pig that you are his leader:

STEP 1 — The Wake-up Call and Attitude Change

The family must stop making excuses for the pig or using "stop-gap" measures (like hiding the pig when guests come) and start taking action. Spoiling Porky must cease. There is a difference between pampering and spoiling.

"Pampering" might be described as buying your piggy satin sheets to sleep on or letting him live in an air-conditioned room.

"Spoiling" is giving in to your pig's demands and letting him train YOU, instead of the reverse. If you think that your pig is becoming spoiled or physically demanding, the entire family must unite and adhere to the game plan. One-hundred percent consistency is required.

Changing your attitude toward your pig is the most important step of all. Realize that the pig is manipulating you and taking control. You must understand that this is because you have ALLOWED him to do this. It is not the pig's fault; he is acting naturally, relying on his herd instinct.

It is time to analyze your everyday interactions with the pig. Is Porky calling the shots? Adjustments must be made.

STEP 2 — Show the Pig You Are in Control, Physically

Pigs often discipline each other with body shoves. It is their social language. If your pig shows overt aggression, shoving him backwards, by applying pressure to his shoulders, can be effective. Pigs seem to respect shoving.

Pushing a pig backward with your leg and saying "NO" also works very well. However, if your pig is very angry, he may not "listen" to you because he is too upset.

A very effective daily exercise to show the pig that you are in control is to gently make the pig retreat by shoving him backwards with your leg, not as a punishment, but as a gentle learning exercise instead.

Again, this is not to punish the pig, but simply to show him that you are capable of making him be submissive. It works well, because when two pigs fight for dominance, the fight is all over when one pig retreats. The pecking order is set, and the pigs are content.

The pig will gently get the message that YOU are in control if you can make him slowly walk backwards.

Start by calling the pig over, when he is calm and had his meal, and use your leg to push him back, just once or twice. Say nothing, but look him in the eye and stand tall. Strong posture is important. After a few hours have passed, call him back and increase the backward retreat to 3 feet. Practice this subtle exercise every day for at least a week. You will be amazed at how it can change your pig's attitude!

If the pig is very large and absolutely will not retreat, a sorting board can be used to make him walk backwards. Don't shove him; the pig will retreat by himself if you get within an inch or two of him. A sorting board is what hog farmers use to herd pigs. It is approximately a 3' x 2' plywood board with a handle to hold onto. In a pinch, a garbage can lid, or any piece of adequately-sized wood, will do.

If the pig growls when you make him retreat, stop and come back later, making the pig retreat just a few

inches. Be gentle, but firm. Do not corner the pig or do anything to make him feel helpless. Using your leg is the most effective way, if it is possible.

STEP 3 — No Free Lunch

Porky will no longer be allowed to wake you up screaming for his meal. He will be called to breakfast when the family is ready to feed him. Begging will get him nowhere and he must perform a trick in order to get a treat. Children will not feed the pig unless they are capable of following this plan. The pig will learn to respect those who feed him and think of YOU as being in control.

STEP 4 — Teach Tricks and Obedience

Teaching the pig to "sit" and do other tricks will display that you are in control, stimulate the pig, and give him some challenges and self-confidence. It will divert his destructive behavior, and he will choose to behave in a manner that earns him rewards. His focus will change. He will be more self-confident after learning to overcome challenges. Start daily training sessions at the same time every day.

STEP 5 — Make Your Pig Worldly

Many pigs have only been out of their yards several times since becoming companions. A confined, bored, spoiled piggy is the most likely candidate to show aggression over his small domain. He has a narrow focus and can be self-centered and territorial. Expand his world by taking him to parks, schools, nursing homes, or pig shows.

Consider getting your pig a buddy
if he behaves aggressively.

House pigs tend to show aggression more often than outdoor pigs.

People love to meet and greet pigs! Make your pig worldly and allow him to have a broader perspective than just living in a small herd. It will dilute his herd instinct.

STEP 6 — Consider Moving Your Pig Outside

Moving the pig outdoors to live may be an option for some people. Pigs tend to appreciate humans more when they don't live in the house with them! Since they don't actually share a common roof, territorialism is rare in outdoor pigs. You will not have as intimate a relationship with your pig, but your pig will immediately show less aggressive tendencies.

STEP 7 — Exercise and Conditioning

A fat pig that barely moves can easily become a spoiled, grumpy, uncomfortable pig. Porkers, like humans, need physical challenges or they may make their own challenges, which could lead to fighting and competing with humans. Portly porkers can learn to go up a few stairs and to take walks around the yard in order to get into shape. Your vet can recommend a proper diet.

STEP 8 — Consider Getting Another Pig

"Only" pigs tend to act more aggressively and territorial than pigs who have porcine pals. This is because the entire focus of a single pig is you, his family (herd). Purchasing a friend for him will give him his own herd to lord over and will provide him with a playmate.

STEP 9 — Establish Boundaries and Routines

Allowing a pig to have the free-run of the entire house can contribute to territorialism. By drawing a "line in the sand" of where Porky can and can't go sends the message that you are in control of the home.

Pig breeders and farmers know that pigs thrive on routine and structure. Change something and the hogs go crazy. Routine gives potbellies a sense of security and imparts that you control their environment.

STEP 10 — Act Like a Leader

Setting limits, being strong, consistent, and firm, but always respectful and kind, go a long way. Never show that you fear your pig; never retreat from him. Your pig will observe you and react accordingly. A happy, well-adjusted pig knows what is expected of him.

Remember that the pig does not really want to be the leader. As soon as you act like the leader, he will be relieved and content to follow your guidance.

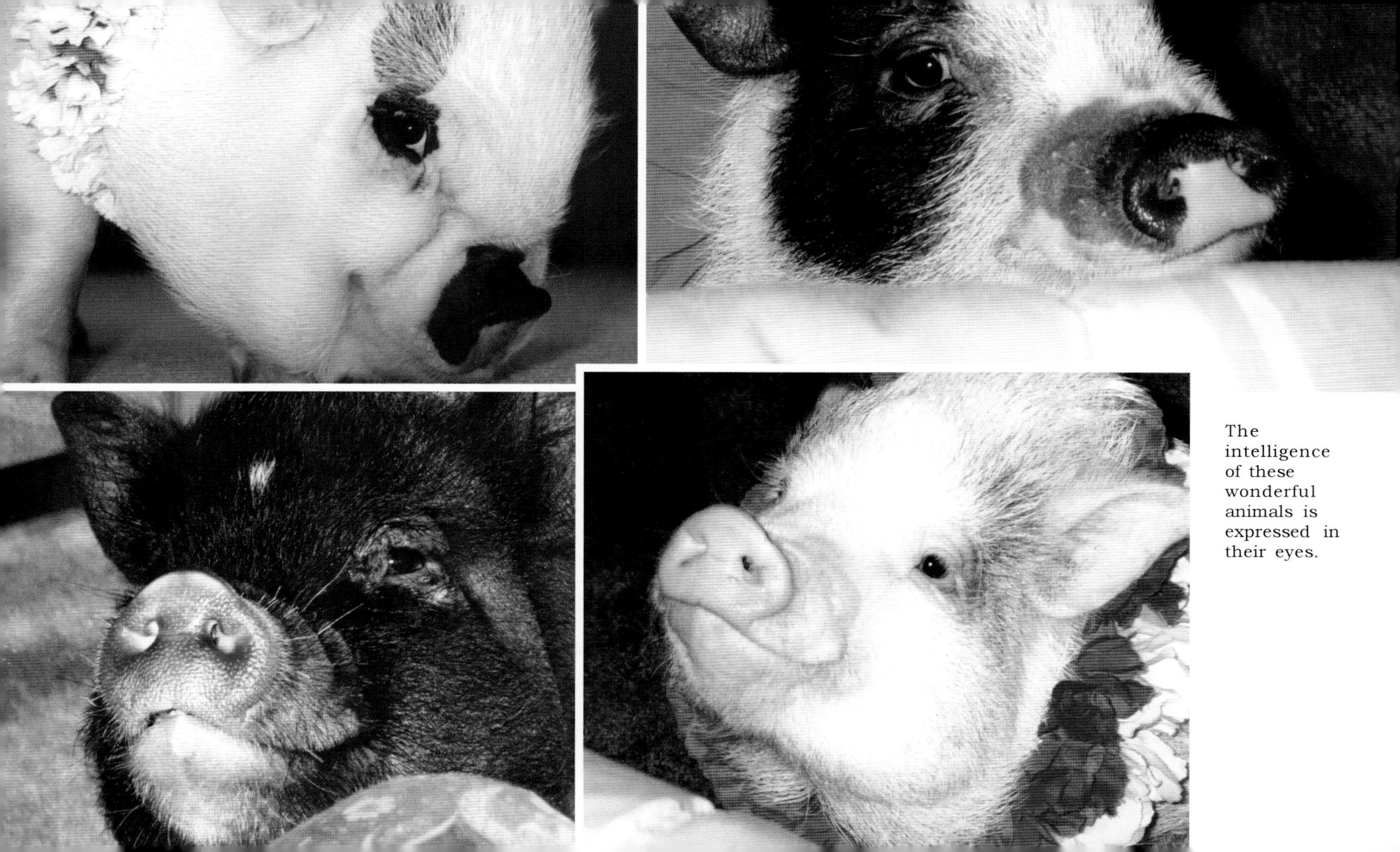

The intelligence of these wonderful animals is expressed in their eyes.

Beginning tricks utilize a pig's natural tendencies, like pushing or rooting. The more advanced tricks go against the pig's natural tendencies and require more time and motivation.

Pigs can
actually be
taught athletic
feats by the
use of positive
reinforcement.

Teaching a pig to perform feats will channel his destructive tendencies in a positive direction.

Whether it be short, medium, or long, a pig's snout is a work of art!

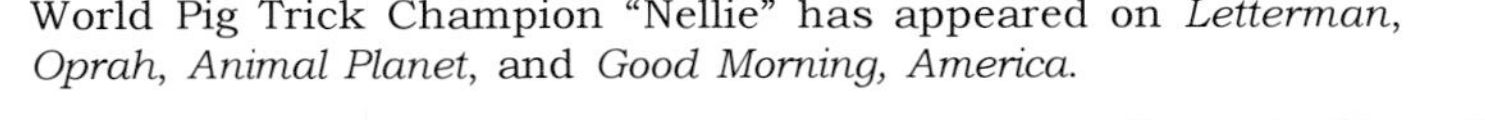

World Pig Trick Champion "Nellie" has appeared on *Letterman*, *Oprah*, *Animal Planet*, and *Good Morning, America*.

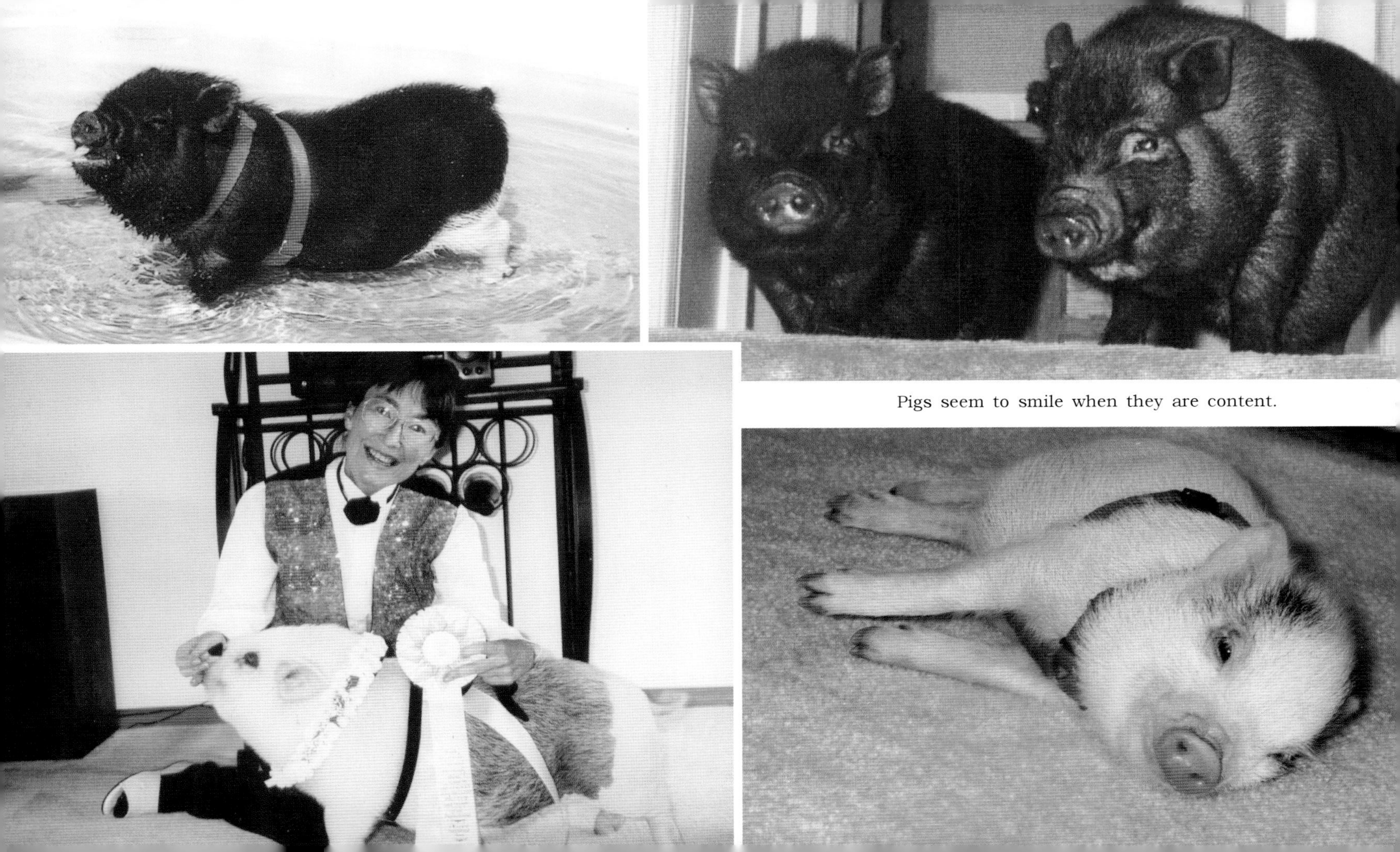

Pigs seem to smile when they are content.

Crowds respect and admire pigs after seeing "Ambassador" Nellie perform!

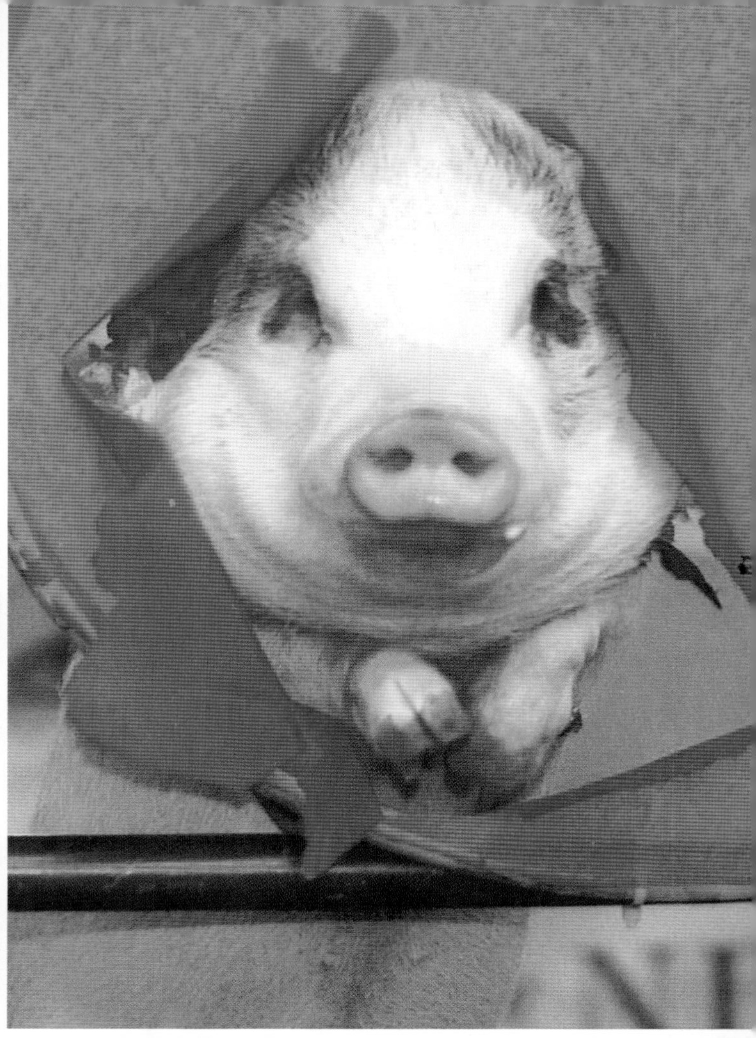

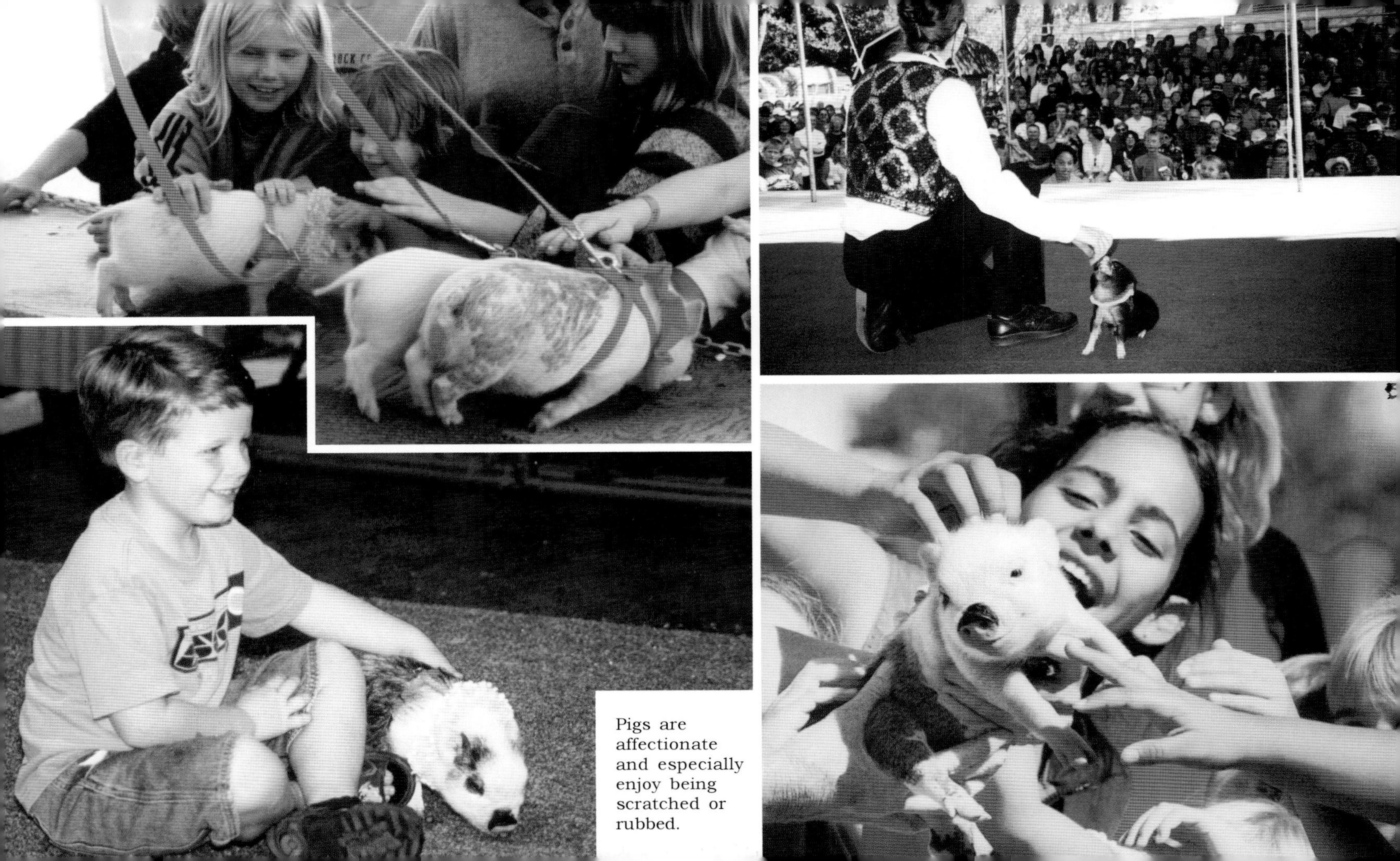

Pigs are affectionate and especially enjoy being scratched or rubbed.

Careful and caring breeders have indoor facilities for farrowing and supervise the birth and socialization of the piglets.

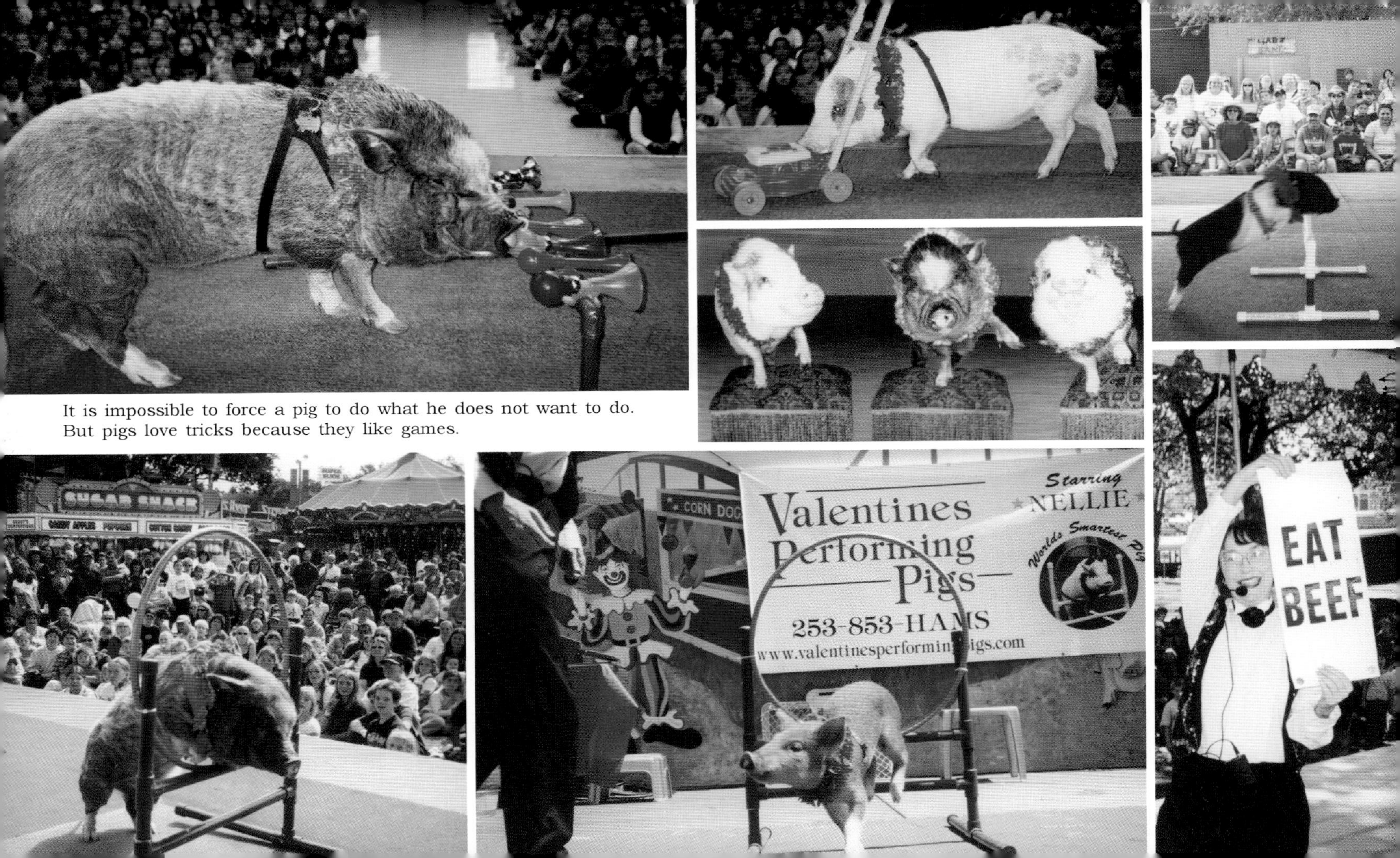

It is impossible to force a pig to do what he does not want to do. But pigs love tricks because they like games.

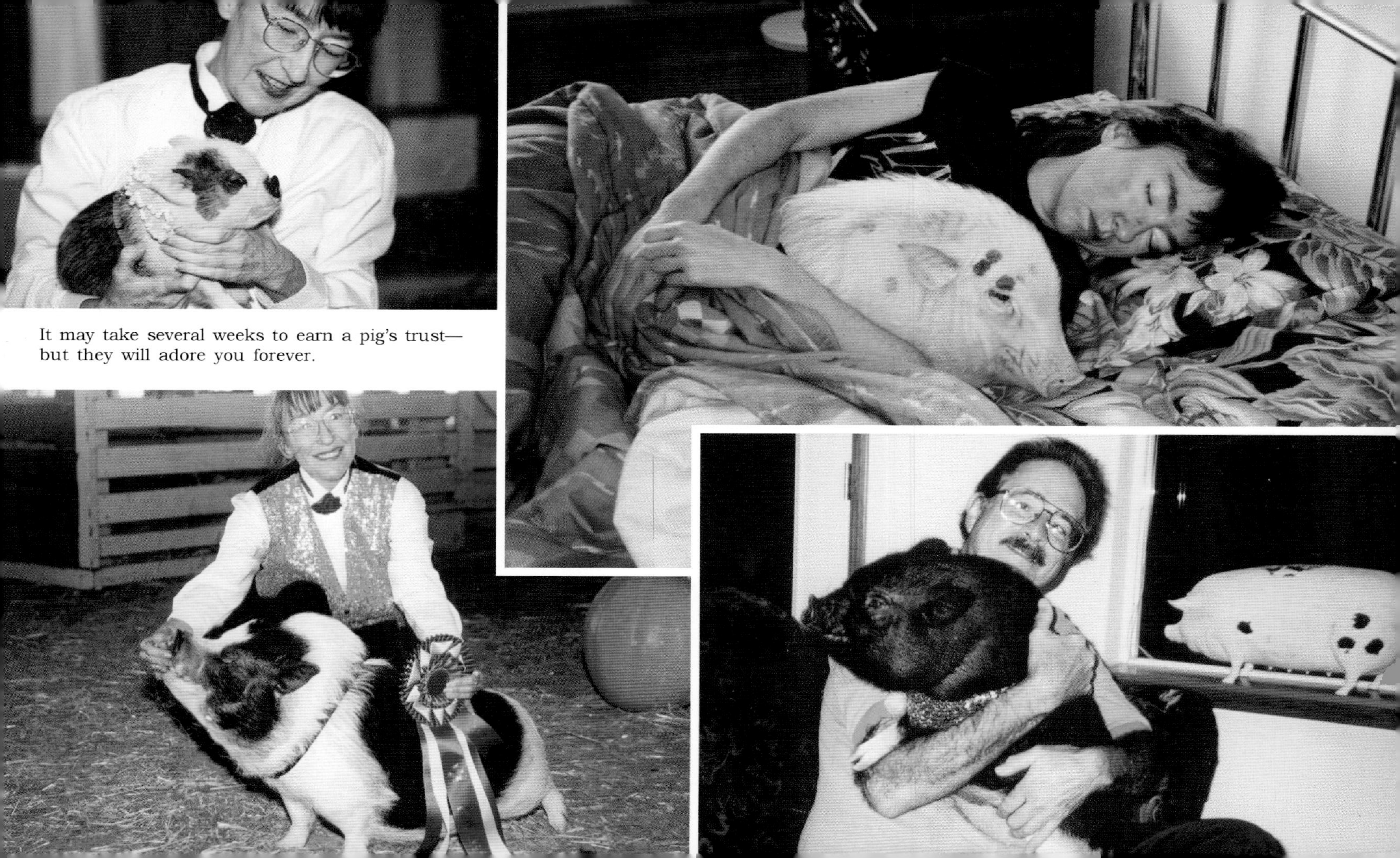

It may take several weeks to earn a pig's trust—but they will adore you forever.

Pigs that are trained to stand on their hind legs stay agile and will probably have less mobility problems as they age.

These pigs are active and fit. Obesity causes pigs to age quickly, lose mobility, and even become blind.

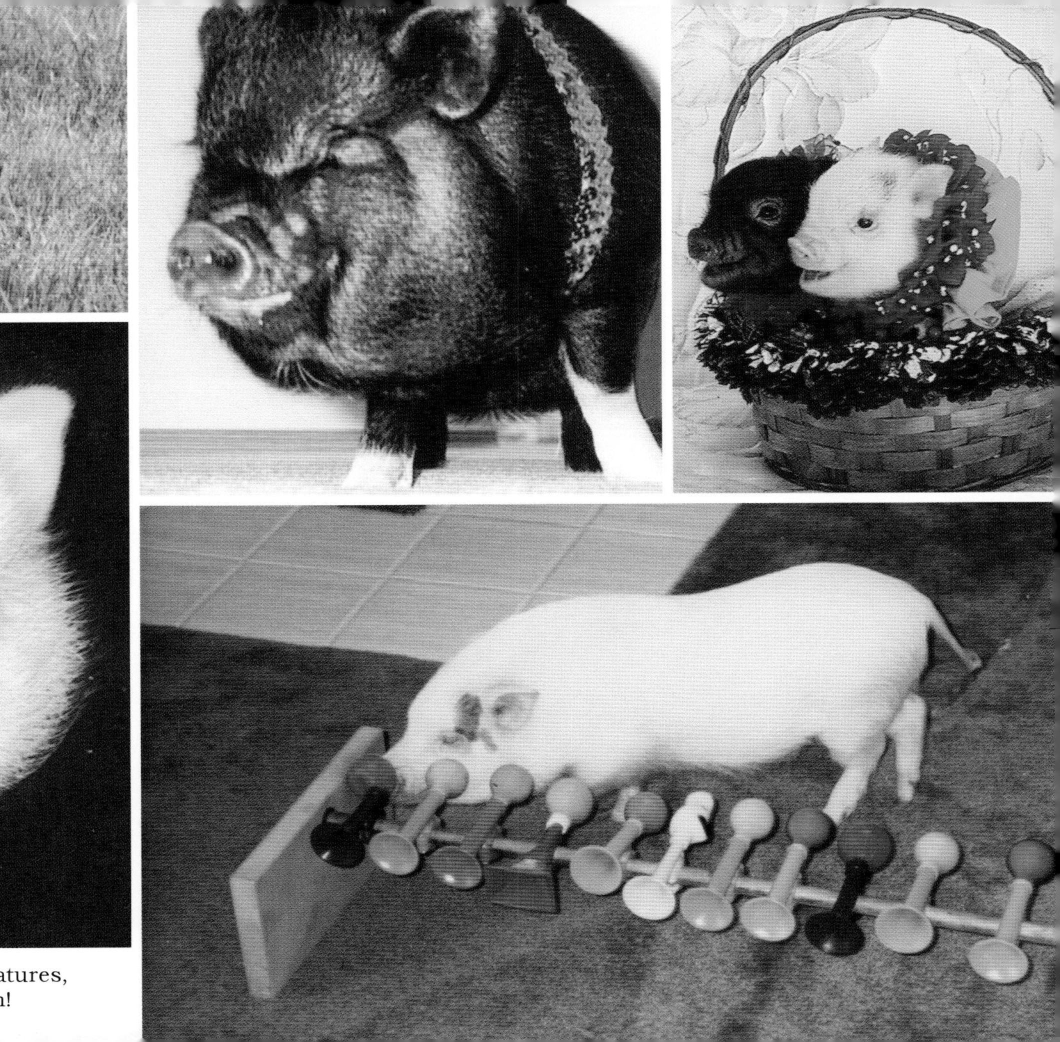

Porkers are hedonists. They enjoy moderate temperatures, a soft bed, and food that is usually not good for them!

Potbellied pigs have many different styles of conformation. Some even wear clothing!

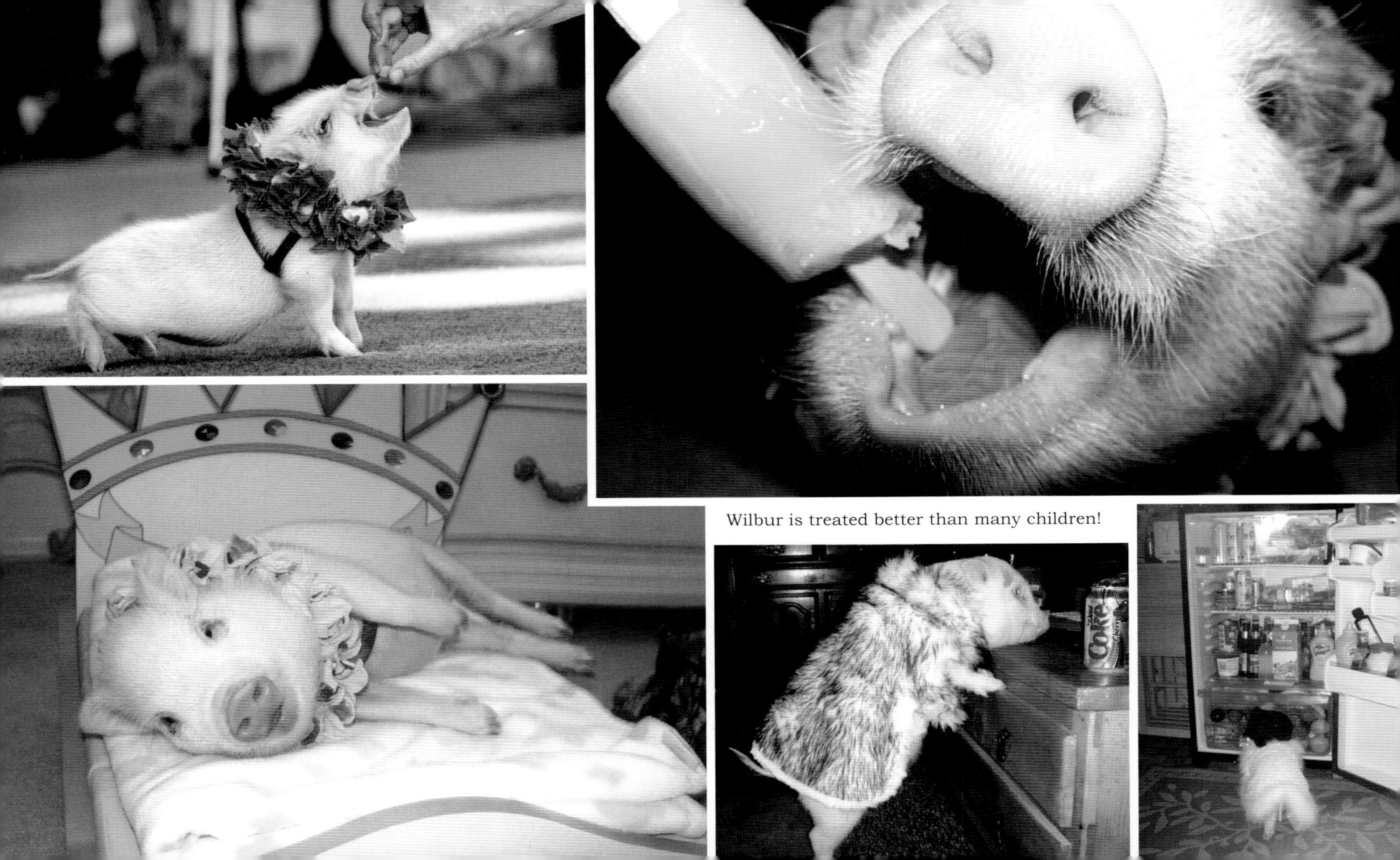

Wilbur is treated better than many children!

Nellie has appeared on *Good Morning America, The Today Show, Late Night with David Letterman, The Tonight Show with Jay Leno, Donahue, Animal Planet,* and more...

Babies only weigh about eight ounces at birth, but grow quickly.

Pigs should be supervised around dogs or young children.

PIGZFLY

Pigs doing tricks always generates huge crowds and teaches the pig to respect you.
See more at www.valentinesperformingpigs.com

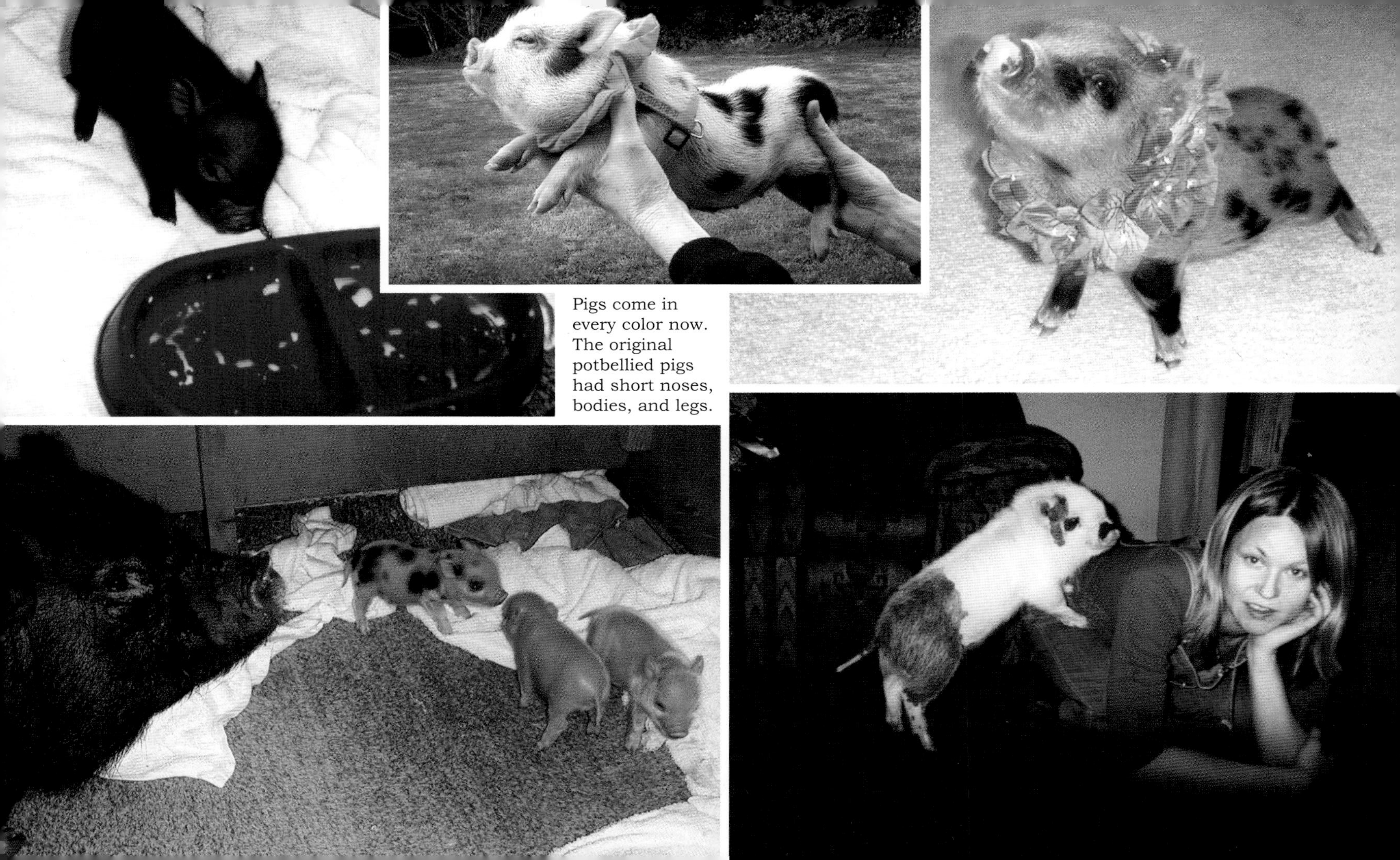

Pigs come in every color now. The original potbellied pigs had short noses, bodies, and legs.

Pigs need time outdoors, even in New York City!

CHAPTER 12

SHAPING YOUR PET PIG'S BEHAVIOR

Shaping your pet pig's behavior patterns will ultimately make him see that you are both his friend and his leader. He will understand that he needs you to make him feel safe and secure, and he will appreciate your home as a wonderful place to be. Teaching your pig to be obedient will impart to him that you have influence over him. He will respect you because he will realize that he *needs* you to depend upon. Pigs love a routine lifestyle and regular training sessions will fulfill that need too. They also need stimulating challenges and something to occupy their time.

Most of all, training your potbelly will shift his focus from destructive habits into positive ones. His home environment will be enriched, and he will become a better pet.

1. USE ONLY POSITIVE REINFORCEMENT

Pigs respond beautifully to positive reinforcement. It makes training a pleasant experience for both trainer and piggy when only rewards are given, rather than punishment. Potbellies do not respond well to negative reinforcement or intimidation of any kind. They tend to

get upset or become "pig-headed." Using a simple method, such as Pig Programming, will make training a breeze.

2. BE CONSISTENT

Pigs are different than dogs. If you are not 100 percent consistent, pigs will take advantage of you. Pigs are rather lazy critters that will always take the shortest route to the finish line.

Because of their intelligence, you must be consistent or your pig will begin to ignore your commands. Keep your specific command words the same as well as your training "style."

3. CORRECT BAD BEHAVIOR IMMEDIATELY

In order for your pig to learn, he must know at once (within a minute or two) if he has done something wrong. If you come home and found that your pig has torn a hole in the carpet, punishing him then will be worthless. The pig will not understand unless he is caught in the act.

4. BE GENTLE

Pigs do not respond well to loud noises or shouting. Porkers are extremely sensitive animals that must look out for their own physical survival. Yelling at pigs upsets them, and they won't listen to angry people. A low, strong, firm voice is always best when you are disciplining or training. If you sound out-of-control, your pig will neither respect you nor trust you. With pigs, aggression begets aggression. Pigs respond well to firm gentleness.

5. BE PATIENT

Pigs must learn in a slow, repetitious manner at first. If your pig is not doing what you think he should be doing in your training sessions, back up. Try to divide the feats or tricks you are training him to do into smaller segments. Teach the pig the first increment before you go on to the second increment. If he is unable to understand what you are trying to teach him, break it down even more.

6. BE A LEADER

Remember, no matter how much you love your pig, he still needs a leader that he can depend upon. Love is not enough! You need to show your pig that you are strong and can take care of him. This can be accomplished by having good, tall posture; talking in a strong, low, predictable voice; and maintaining good eye contact. If the pig sees you as a leader in the family herd, he will respect you and want to please you.

7. BE LOVING

Spend lots of time with your pig and teach him new challenges. Take him to intriguing places, and allow him to have fresh experiences. Be kind and give him attention so he won't demand it by being destructive. Giving him lots of belly rubs and praise will help make him a good pet.

8. DON'T OVER-REWARD

Pigs appreciate even one paltry kernel of popcorn! They deserve to be paid for honoring a command. The

How you treat and communicate with your pig will determine if he makes a good pet or not.

reward should be swift but small. And a pig should never be paid for "half-doing" any feat if he's already learned how to do it. Demand that the pig earn his reward. Watch his weight so he can continue to lead a quality life. An owner should not become a vending machine.

9. PRAISE YOUR PIG

Potbellies love to be rewarded with food, but they love a pat on the head too. Many pig owners rely only on food to convey their pleasure. This is cheating the pig. They appreciate your enthusiasm and encouraging words also.

10. TALK TO YOUR PIG

It is amazing how many actual words and word commands the clever pigs can learn. Verbally communicating to them establishes a bonding and a rapport between the caretakers and the potbellies.

11. PUT YOUR PIG ON A ROUTINE

Structure makes your pigger feel secure and content. Set up training sessions at the same time every day. Your pig will look forward to it. Feed him and tuck him in at approximately the same time every day as well.

12. RESPECT YOUR PIG

Pigs demand our respect. Pushing or shoving your pig, scaring your pig, or not being sensitive to your pig will set back his training. Pigs have very low tolerances for being physically manipulated. Instead of bullying your pig, try to motivate him.

13. LEARN TO THINK LIKE A PIG

Potbellies usually have a rational reason behind most of their actions, even though it might not seem like it at the time! Taking the time to study your pig's behavioral patterns and observing his more subtle behaviors will allow you to understand why he does what he does. Once you understand the reasoning behind his actions, you will be in a position to better shape his future behavior.

14. LET A PIG BE A PIG

Training and shaping a pig's behavior can be very beneficial to the porker. But attempting to change a pig into a dog or any other animal is impossible. Learn to appreciate the uniqueness of the Vietnamese potbellied pig. Natural habits, like rooting, can be dealt with in a constructive manner.

15. AVOID DISTRESSING YOUR POTBELLY

Potbellies react poorly to stress, and it can even be life threatening to them. Usually there are ways around allowing a pig to become distressed. It just takes planning ahead and patience. In the end, both the owner and pig will benefit.

Pig Behavioral Problems from A to Z

Aggression Toward Boyfriend

"My pig charges my boyfriend. We are getting married soon. How can I make Ham like him?"

Ham is exhibiting aggressive behavior because he wants to dominate your fiancé. Pigs have a pecking order. When confused about their social ranking, they will challenge the most vulnerable people — children and houseguests.

Your first impulse might be to put the pig in another room when your boyfriend visits. Do not do this! This will only encourage the pig to try to dominate more humans if he wins the battle. Instead, after your marriage, have your boyfriend become the pig's sole caretaker. He will provide all food, affection, and training. If he makes the pig sit, it will impart to the pig that he is the pig's leader.

All other family members must ignore the pig. The porker must earn all praise and food from the new husband. This will establish your husband as a leader in Ham's eyes.

In order to put an end to the charging now, the pig should never be allowed to achieve dominant status over your boyfriend or anyone else. Your boyfriend must win every confrontation they have. When the pig charges, have your boyfriend firmly make Ham retreat by pushing him backwards with his leg. This is an exercise, *not* a punishment. In addition, you will need to follow the Ten-Step Program in the book for halting aggression.

Once a pig has reached dominant status, punishment cannot be used to correct his hostile actions. This is because the pig is doing what comes naturally. He is trying to lead his herd. Punishment will only make him angry and more confused.

If your boyfriend is consistent in making the pig retreat, using the pushing exercise, and training him, the pig should cease all aggressive behaviors toward him. Ham and your boyfriend will become buddies.

Begging Porker

“Snort is always begging for food. When I sit down on the sofa, hoping to enjoy a few potato chips, he stands on his hind legs trying to get the whole bag! At dinner, he almost jumps on my lap, hoping for a morsel. I give him one, hoping that he will be satisfied and go away. He just stands there, begging for more. My three-year-old son gave him a piece of candy and he knocked him over while trying to get more. What's a person to do?”

Well, it's very apparent where this species got the name PIG. Pigs live to eat, and if these adorable critters ever get their ample bellies full, they sure don't tell us about it!

Give a pig an inch, and he will take 100 miles, especially if it is anything that fits into his mouth. Snort needs to learn manners.

The secret to preventing begging is for you to always be in control of dispensing food to the porker. If the pig roars up to you demanding food and you give in, the PIG is in control. He has trained you. From then on, you have created a porcine monster. You are no longer a person to respect. To him, you are a free vending machine! And he will do whatever he can to pull the handle without paying.

It is never too late to start teaching a potbelly good habits. All family members must agree not to feed the pig while they are eating at the table or snacking. They

must stick to the plan 100 percent or they will train the pig to be PERSISTENT in his pesky begging and pestering.

It sounds simple and it works very well, but only if you can stick to the plan. I know it's hard to resist a "starving," adorable, potbellied pig, with his little hoofers up and a grin under his snout, asking for just a tiny snack . . . but it must be done.

At first, the pig will be indignant, but after a few days, he will find better things to entertain himself. Just ignoring him works better than chastising or pushing him away. If you try and push him away, the pigger will make a game of it. And pigs love games! So, instead, when Snort approaches, do not even acknowledge his presence as you are eating. Do not look down; do not say a word. Pretend you don't even know he's there, and be consistent. You must be careful not to spill any food or you have defeated the purpose. Feed Snort in a special area away from the table before you eat your meals. Continue ignoring him if you are eating. If you must feed him at times other than official meal times, make him earn it by performing a trick.

It is best if your three-year-old does not hand-feed the pig. Otherwise, the pig can become quite rough in attempting to get treats. The toddler is not old enough to stand up for himself against the pig. It's just a matter of time until the pig senses that and might try to establish dominance over your son.

A pig that is never allowed food while you are eating will not beg.

Biting Shelter Pig

> "We just got in a new pig — an older barrow. He tried to charge and bite me! What do we do?"

He is biting out of fear. He may have been abused, and he is frightened and wants to protect himself. The pig doesn't trust humans because he has been hurt by them and may be hurt again. He is trying to warn you to back off. He is confused and doesn't understand why he is being jostled around. Pigs hate disruption. The frustrated barrow is communicating in a natural manner that he is uncomfortable and leery of humans. An intelligent animal will act this way in order to survive. Pigs possess little in the way of body defense and tend to remember abuse for a long time. It's hard to put the blame on the pig. However, this behavior must be modified.

Leave him alone for a few days and just feed him, no verbal contact. Then gradually begin to earn his trust, just as you would a new piglet (see piglet training section). Just be sure to protect yourself from getting hurt.

Put him on a regimen, and be very gentle and sensitive to his behavior. Learn to read his signals. Pigs almost always have a good reason for acting the way they do. It's up to the caretaker to first understand why the pig is biting, protect himself, and then use "swinecology" on the pig.

Always have a very fast way to get out of his pen if

there is trouble. He may be depressed because his owner has abandoned him. Give him plenty of space, time, and love.

If he still charges in a few months and his biting appears to be territorial, try displacing him with your knee when he is in a gentle mood. Talk to him in a pleasing manner as you force him to inch backwards. Start with only a few inches and increase the distance slowly over a period of days. He will recognize you as a herd leader and act respectfully toward you if you can make him physically retreat. But this must be done with extreme gentleness or it will cause anger in the pig.

If you get down low, you are more vulnerable to being attacked by the pig if he is so inclined. Stand tall! Your pig may be reacting to years of abuse by his former caretakers. If you are compassionate, but show yourself to be a strong leader, you will earn your pig's love and respect.

A happy, normal, potbellied pig does not bite or charge.

Blocking Doorways

"Wilbur is always blocking doorways or entrances. Sometimes he sleeps on the front porch steps, and I'm tired of walking over him. He's very grumpy and demanding. What should I do?"

Wilbur is getting demanding because he is taking over as Top Hog. By stepping over Wilbur, you are sending

him a message that he is in control of you and your home. Instead, have Wilbur move. In the animal kingdom, the lower animals in the social hierarchy are forced to move and accommodate the higher animals. In order to establish yourself as Wilbur's leader, gently use your legs to push Wilbur aside. Carefully knock him off his "perch." At first, use happy talk while you are doing this or he may snap at you.

Wilbur should also be taught to sit and stay. Learning to be patient will make him a better pet that respects your space. He will not be underfoot all the time.

These exercises will go a long way. The physical displacement of Wilbur will make him respect you, and he will be a happier pig. Pigs want to have a leader to look up to and guide them.

Chewing Household Items

"When I came home from work, Hamlet had destroyed a plant and chewed up a pair of my husband's leather shoes. I yelled at Hamlet and now he's pouting. What can I do to make him stop wrecking things?"

Pigs get bored easily, and they love to explore and forage. The odds are that Hamlet just wanted something to do to pass the time. Pigs are extremely intelligent critters that need something to occupy themselves with. If pet owners don't provide something acceptable for them to do, their only choice is to get into mischief.

Perhaps you are giving Hamlet too much freedom. He may view your home as an amusement park. You must teach Hamlet which objects he can play with and which ones he can't WHILE YOU ARE AT HOME. It is important to be there to supervise and set patterns.

As Top Hog, you must reprimand Hamlet strongly if he plays with forbidden items when you are in the house. Tell him, "NO, leave it," and briskly lead him away from the item. Then, give him an acceptable toy to play with and praise him if he takes to it. Be consistent.

Buy Hamlet a Manna Ball filled with goodies. These large, round, plastic balls slowly distribute pellets of pig feed or popcorn on the floor, as the porker roots them along with his snout. A rooting box (a plastic box filled with river rocks and birdseed) is an ideal acceptable toy. Pigs love to shred yesterday's news, discarded telephone books, and old magazines. Plastic bowling ball sets seem to fascinate pigs also. They will spend hours knocking over the pins with their noses, and then pushing them along the floor. Large, paper, grocery bags and safe, chew toys are also a big hit with the porcine set.

The next time you come home and Hamlet has chewed on something, don't bother to punish him. You will only confuse him, and he will think you are crazy. It is not fair discipline to the unfortunate pig. You must catch Hamlet in the act before it is productive to respond to the action.

Most pig owners do pig-proof their houses to some extent. Pigs are curious animals that desire to open every cupboard, looking for fun or food. Dangerous toxic items must be stashed well out of the porksters' reach. Such items as dog and cat food, cereal, candy, snacks, etc. must be locked up unless you want to tempt fate!

Hamlet could be put in a child's play yard or in a pig-proofed, separate room while you are out. But he should be provided with lots of fun toys for something to do.

Exercising Hamlet before you go out and practicing a few tricks with him would satisfy some of his need to be stimulated and perhaps even tire him out a bit. Letting him play outdoors will help.

Dog and Pig Sparring

> "My mild-mannered German Shepherd and Piggy Sue do not get along all the time. What should I do?"

Pigs and dogs both have pecking orders in their herds or packs. If there's no clear leader between them, this can cause problems with feuding. If the dog and pig have just recently met, you can let THEM decide who is in charge. Introduce them to each other gradually and closely monitor them in order to see that neither one gets hurt.

Otherwise, if the animals cannot decide which one is the leader, their behavior must be carefully supervised.

Pigs are not physically capable of defending themselves against big dogs. Some piggys are maimed or killed by the family's beloved pet dog. Prior to this type of incident, the dog doing the attacking is described as sweet and docile by their devastated owners. It is a complete surprise to the owners!

Unfortunately, dogs are predators and have an instinct to prey on animals like pigs. Pigs are prey. And grumpy pigs can actually instigate an attack by teasing the dog or taking his food. The dog will respond very harshly and the fight can get out of control.

The porker will always lose the fight.

If you witness a fight, break them up by using a spray bottle and take them to separate rooms of the house while chastising them.

Make sure that the dog and pig both have proper obedience training, so they will accept your leadership and your directions.

Both animals should be neutered/spayed. This will cut down on aggression and make them more trainable.

It is a good idea to always feed them separately, and never leave any food out that they might fight over.

But the bottom line is that an unsupervised large dog and a potbellied pig should never be left alone together. If you are unable to supervise them, it would be best to find a new home for one of them.

Eating—Refusing to Eat Pig Food

"Penelope won't eat pig pellets! She's been raised on dog food since we purchased her as a piglet. When we try giving her potbellied pig feed, she just dumps the bowl and leaves it. So we still feed her dog food. How can we make her like pig feed?"

Dog chow is not good for pigs. It's too high in protein and calories for swine to live on. Potbellied feed, on the other hand, is low in protein and fat, and high in fiber. Scientists who studied the nutritional needs of potbellies formulated it. They created a feed that will enable pigs to lead long, healthy lives.

Penelope enjoys the taste of dog food because it is higher in salt and fat. Pigs actually have discriminating tastes. Just like humans, they prefer salty, fatty, and sugary food to what's good for them.

Potbellies are just like spoiled little kids who want to eat desert before dinner or crave candy all day long. If a pig can con you into buying him very palatable, calorie-laden feed over the rather mundane-tasting pig food, he will!

A pig that lives on dog food or eats too much will probably have impaired movements, back problems, and a belly that is too rotund. Start teaching Penelope to develop a taste for pig food by mixing in just a few pellets in her bowl. If you put in too many, the devious little pig will probably sort them out separately and reject them. If she does sort them out, water down her food and mash it all together so that she can't separate the TWO kinds of feed.

Gradually increase the amount of pig feed. She may be pig-headed, but she has to survive. In no time, she will be squealing for pig food.

This works well because it gives her time to adjust to the new, more subtle taste. Eventually, she will be weaned off the canine food.

Eating—Eating Too Much

"My porker, Porkchop, overeats. He acts like he is always starving, and he's getting way too fat! What can I do?"

In Vietnam, potbellied pigs consumed large amounts of water plants that grew in rice paddies. They dined on water hyacinths, water lentils, sweet-potato foliage, and rice waste with a few snails thrown in! This diet is tremendously high in fiber and roughage. It is suspected that the pigs' pendulous trademark belly evolved from the great amount of bulky roughage consumed by the porker. The pigs were said to have indulged in an early morning vegetable gorge, slept the muggy afternoon away, and pigged out again in the cool evening. To an Oriental

pig, he is not through eating until his stomach feels very, very full because of this gorging instinct.

If Porkchop didn't get fed enough food when he was young, he will have even more of an insatiable appetite than a normal potbelly. Pigs will definitely eat, even when they are not hungry. Usually the only time they turn their snouts up at vittles is when they are very sick!

Taking your pig to a potbellied pig veterinarian as a precaution is a good idea. Ask him to appraise Porkchop's body conditioning, and try to accurately describe to the vet exactly how much food Porkchop is consuming. Your vet can tell you if he is over-conditioned and the proper amount of food that he should have. He can suggest a proper diet too. Serve it in two meals a day: one in the morning and one in the evening.

Obesity is a major health risk in potbellied pigs. They are prone to permanent lameness and arthritis if they are carrying too much weight on those inadequate, skinny, short legs. And the porkers get all the same heart complications that humans do if they are too portly. As the pigs age, it is even more important to watch their weight. Once they do get fat, it's hard to take it off because the porkers become inactive and grumpy, and exercise can be painful to them. Prevention is the key. Weighing your pig periodically is a good idea too. Keep on top of it. An obese pet pig does not lead a quality life.

Potbellies should only eat feed that was specifically formulated for them by scientists. Dog food is too high in protein, fat, and salt for a potbellied pig. Commercial hog food may not have the proper nutrients and may be medicated. Potbellied pigs should not have medicated feed. Commercial hog food is usually intended for pigs

that live on it only for a very short time. You want a feed for Porkchop that gives him the nutrients to lead a long, healthy life. Potbellies can live for over 15 years, and they need a diet that is low in fat and protein, and high in natural fiber.

Do not give in to Porkchop's begging, even if he looks up to you with his pleading eyes and porcine grin. Don't feel guilty. You are doing Porkchop a favor by showing him that you care about his welfare. When he does earn a treat, give him chopped up carrots, broccoli, and green leafy veggies. These bulky vegetables will puff up his tummy and give him the sensation of being full. Air-popped, unsalted popcorn is appreciated by porksters everywhere, is low in calorie, and provides plenty of fiber.

Exercising a potbellied pig can be a lot of cheap entertainment! Hook Porkchop up to a leash and harness, take him to the park, and chat with the crowd that will undoubtedly form. Start slowly, because if the pig is obese, it will take more effort at first. Make it an enjoyable experience for Porkchop with plenty of happy talk and a few grapes along the way!

If the weather is hot, do it on another day. Potbellies are prone to heat stress. Another way to encourage Porkchop to painlessly work off a little lard is to broadcast his food pellets on your patio or outdoor area at mealtime. He will have a ball waddling around sniffing for them and turning objects over. And if you do have a patio, the action of his hooves rubbing against the concrete as he searches for pellets will help keep them short!

About three times a week, place a special treat (such as a handful of carrots) in the farthest corner of the lawn and let Porkchop discover it. Pigs love the surprise of

unearthing special goodies. It's even more fun if it isn't a sure thing every day.

A quality, potbellied pig food will supply all the essential vitamins and minerals that Porkchop requires. Feeding such well-known and established brand names as Ross Mill's Champions Choice, Heartland, and Purina Mill's Mazuri will make sure that Porkchop receives more than adequate nutrition.

Try to separate your own need to feed the pig from the pig's need to be healthy and live a long, happy life. Ask yourself: "Am I feeding Porkchop this fattening treat because I enjoy doing it for the fun of it, or am I doing it for Porkchop's future?"

Fear of Taking a Bath

"Penny is scared to death to take a bath. We took her in the bathroom, kicking and screaming, and it was not a pleasant experience for any of us. How can we make it less of an ordeal?"

Penny is frightened of taking a bath because potbellies have a natural aversion to bathrooms and slippery, wet surfaces. She also is afraid that the water may be too hot or too cold. Pigs don't like the feel of water (except to cool themselves off in very hot weather). Getting in and out of a shower or bath is difficult for a stubby-legged, chubby porker too!

Gradually, start getting Penny used to just being in the bathroom with no negative experiences. Even if you have to feed her a whole bowl full of food, coax her to come into the bathroom again, despite the last unpleasant experience. When she does, praise her and entertain her for a few minutes. Asking her to sit or spin will amuse her, and she will realize that bathrooms aren't just for bathing. Get down on her level and scratch her tummy and tell her how much she means to you. Repeat this every couple of days until she thinks of the bathroom as a happy place to be.

Cover the floor of the bathroom with plenty of soft, non-skid rugs, so that Penny won't have to fret about falling and injuring herself. After Penny is used to the room, turn on the shower or bath faucet, allowing her to just hear the water. Give her a treat when it's turned on. Turn it off and back on again, and pay her once more. Every time the water is turned on, she gets a small morsel. The sound of running water will soon be music to her ears, and her mouth will start foaming in anticipation of a goody being dispensed. This is a true "conditioned response."

Start coaxing her into the shower stall with grapes until she is comfortable going in and out of it on her own. This may take several days. Gradually, add a tiny amount of water to the stall and increase the depth in it each time she goes in the stall for the grape cache. Make sure that the water temperature is not too hot or too cold. Say "bath" every time she gets her hooves wet.

As she adjusts to having her feet in the lukewarm water, start gently massaging her back with wet hands. Introduce her to a mild baby shampoo by letting her sniff

it first. Use very little pressure from your hands and try not to frighten Penny. Her bath should be a two-person job: one person to feed and reassure Penny, and the other to handle the water and soap and actually give the bath. In order to motivate Penny to like baths, add grapes to the water and start gently lathering her. Reassure her with gentle, soft-spoken words and continue talking to her. Keep steadily adding treats as you work.

After the bath is done, have some hot towels handy (fresh from the dryer) to massage her. She will enjoy the warmth and remember the pleasurable experience. If you must use a tub to bathe Penny, make sure you line the bottom with towels and mats so that she doesn't panic when she finds out she has no traction. If she attempts to jump out, ask her to "sit" and reward her. The first time you put a pig in the tub, don't add water. Make it a "dry run" for the nervous pig that is going to feel a little confined at first.

Taking the time to gradually get Penny used to the bathroom will put her fears to rest, and make her feel safe and secure. She will realize that a bath can be very fulfilling, in more ways than one. When it's time for her next beauty treatment, say "bath" and watch her run toward the shower stall!

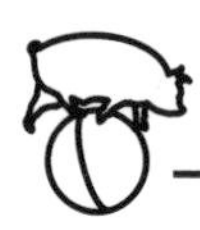

Foaming at the Mouth and Trimming Tusks

"Roscoe, our 6-month-old neutered male, is foaming at the mouth all the time. Is this normal? And when should I get his tusks taken care of?"

It is normal for potbellies to foam at the mouth when excited and happy. Pet pigs foam at the mouth when they anticipate a meal or for no particular reason at all. It's just part of being a pig. The frothy stuff can be as thick as shaving cream when they are all worked up over something! It does not signify that a porker has rabies.

Roscoe should not need his 4 tusks (canine teeth) trimmed until they start protruding from his lips and causing problems. Roscoe's tusks may inadvertently scrape your legs as you walk by. Tusks can damage walls or furniture too. The 4 tusks will continue to grow throughout Roscoe's life and will probably need to be trimmed periodically so that they don't cause damage to himself or you. These canine teeth can potentially damage Roscoe's own mouth if they grow at improper angles or if they curve inward. In the wild, male pigs use their tusks to defend themselves.

Boars (intact males) grow longer tusks than barrows (neutered males) because the tusk size and length are related to the amount of testosterone in the pigs' bodies. The permanent tusks will erupt at about 10-15 months of age.

The tusks should be cut, rather than removed, because they are attached to the pig's jaw. Removing the tusks entirely could fracture the jaw or compromise it. It also leaves a hole that can easily get infected. Therefore, a veterinarian should do this procedure.

Females have tusks too, but they grow much slower and probably will not ever need to be trimmed.

If Roscoe's tusks do not present a problem, you don't need to get them trimmed. Just monitor his mouth for tusk abrasions that could result in infections.

Grinding Teeth

> "My pig, Henry, grinds his teeth. It's annoying. What can we do to make him stop? The sound of it drives me up a wall."

Yes, it can be annoying!

Teeth grinding in pigs sounds like rocks being ground against crushed glass. Potbellies grind their teeth loudly together for a variety of reasons. Henry could be teething if he is less than 2 years of age. He could be a little distressed. Or, most likely, he is a very content potbelly who enjoys a little tooth massaging after meals and naps!

In the wild, pigs grind their canine teeth (tusks) together in order to sharpen them. Most potbellies seem to grind their teeth . . . just because they are PIGS.

It comes with the territory, although some pigs seem to be professional grinders.

There is little we can do about it, except learn to enjoy it as part of sharing life with a potbelly. Sometimes, providing the pig with a safe chew toy or giving your porker something to do will help.

Just be glad that pigs don't bark like dogs. It's a much more offensive noise.

Hooves—She Won't Let Me Trim Them

"Precious won't let me trim her hooves. We have goats and it's no problem with them. What is wrong with her?"

Potbellied pigs have virtually no defense against predators except to run.

They like to keep their hooves on the ground.

In addition, they don't like to be restrained nor have parts of their bodies grabbed until complete trust is established. The only time that their body is restrained is when an animal is carrying them off to eat them. Once trust is established and the pigs feel safe and secure, costly hoof trimming may be done by the owner.

Precious is leery of your handling her feet. There are several ways to overcome this. Pleasing the porker by giving her a gentle belly rub might enable you to slowly start desensitizing the pig to having her hooves fiddled with. As with most pig training, this must be accomplished in baby steps. If the pig revolts, you are going too fast and must back up.

Every day, give Precious her usual belly rub, but start stroking her feet first. This will make her realize that if someone is massaging her legs and feet, even better things are to come. Be consistent, talk to her, and start touching her hooves and applying pressure. Get her used to this for at least several days. Do not reward her with food, or she will have a tendency to jump up in anticipation.

After Precious is used to your handling her hooves, introduce the hoof trimmers to her. Let her sniff them and get used to the sounds they make as they open and close. As you give her the routine belly massage, open and close the clippers many times. This will desensitize her to the funny noise that they make and make her associate the noise with pleasurable experiences.

Finally, you are ready to start. Take only one snip from a hoof, then stop. Start again the next night. If the pig is concerned, you are going too fast. Every night, increase the amount of trimming you do.

For a pig that absolutely REFUSES to let you actually hold the hoof, you may trim the hooves as he stands on them and eats from a bowl. Talk soothingly to the pigger. Place a bowl of feed on a thick carpet sample and as the porkster is chowing down, clip a hoof, taking tiny slices at first, while the pig is bearing weight on it. Place one blade of the trimmer under the target hoof. The thick carpet will have "give," allowing you to accomplish this feat of getting the blade underneath. If the pig starts to balk after a few seconds, back up, and start again tomorrow. This is not an ideal way to trim hooves because it's difficult to level the sole plane. But it will help a pig that has overgrown hooves and needs immediate help.

It is important that whenever you handle your pet pig, you do nothing to set back the trust and confidence that your pig has in you. With pigs, you must EARN their trust. Causing the pig to squeal, struggle, or fear you may set back a relationship that has taken months or even years to build. Be careful not to cut into the quick of the nail and hurt the pig. Pigs do not forget; pigs do not forgive. Taking a few extra days to desensitize Precious to hoof trimming, instead of forcing it on her, will ensure that your close bonding has not been damaged.

Hooves are easier to trim after the pig has taken a bath or stood in a wading pool. It softens them, and the job can be done quicker.

Housing

"We are getting a potbellied pig soon. What kind of living arrangements are best? We have heard differing opinions and are confused."

When the potbellied pigs first were imported into the U.S. in the 1980s, they were marketed as "condo" pets. They were described as "yuppie puppies," and many people did house and treat them as miniature poodles with bristles and hooves!

Since then, we have learned much about what makes the pigs content and secure in their household environment. Pigs are not dogs or cats, and their needs

are different. In addition to the basic needs that all animals have, pigs need certain types of environmental enhancements in order to be happy hogs. Pigs need to graze and root a little, pigs need to lie in the sun, pigs need to explore, and pigs need new and stimulating experiences and challenges.

A pig's sense of smell is very powerful, and their snout is their major sensory organ. They need to indulge in some form of nibbling or foraging as well as playful nudging. They need something to DO, and they need to use their minds.

If all of these specific needs are not met, then something else that enriches the pig's life must be substituted or the pig may not make a good pet in the eyes of his caretaker. An unhappy, bored pig may turn into a belligerent pig. Or he may become destructive around the house.

The easiest way to fulfill the pig's foraging and exploring needs is to make sure that the pig has time outdoors in order to get time to do what pigs naturally do: sniff around, uproot objects, knock things over, and get his nose stopper dirty! If these needs are NOT fulfilled outdoors, the pig will tend to root and chew on things indoors, where pet owners don't want him to. (Certainly the pig can be trained to root and play in an acceptable place indoors, however, only if the owner is willing to take the time and effort to do so.)

An indoor-outdoor living arrangement works well for many potbelly owners. A dog door can be installed, and the pig can choose where he wants to be. Or the pig can be let outside for a certain amount of time every day to graze and forage. In very hot or cold weather, the pig can take advantage of the more moderate temperatures inside the home, using the dog door, and always be comfortable.

Many pigs live totally outdoors as beloved pets. In very hot climates, the pigs will need wading pools, plenty of fresh water, and protection from sunburn in the form of shade. In cold climates, they need a doghouse or other shelter that protects them from drafts and moisture. Dry, clean bedding can be made up of blankets, hay, or straw. Pigs in extremely cold climates may need a safely mounted heat lamp. All pigs that live outdoors need to be protected from dog attacks or wandering off by a strong fence. If your pig lives outdoors, you may have to work harder than an owner who has a pig living indoors does to achieve intimacy and bonding.

Some people have pigs that live indoors virtually all the time. If this is the case, it is probably going to be a struggle and a challenge to entertain the pig and give him something to do all day that isn't destructive to the house or otherwise intolerable. It may be an uphill battle, but it is possible, for some. Using various diversions, such as rooting boxes, Manna Balls, children's toys, safe chew-toys, and training, will help.

Taking the pig on numerous trips and outings will enrich his life. This type of housed pig will require lots of good quantity time, as well as quality time, spent with the caretaker. The potbelly will need plenty of attention and some creativity from you. Having, at the very least, a medium-sized yard for the pig is a must or taking the pig on long walks daily to a park.

This living situation will take more of a commitment from the owner than it would for pigs that have more outdoor time.

Pigs have the potential to make good pets. But the caretaker must realize what the pig's basic needs are, and make sacrifices and adjustments according to the needs of the individual pig and his circumstances.

Pigs need to be pigs.

Integrating a New Pig

> "We are going to buy a new pig, and our existing pig, Sweet Pea, has always been a spoiled housepig. I'm not sure that she is going to welcome a new porkster into the family! Is there anything we can do to make it easier on her?"

Some potbellies will welcome a new playmate, and others are not happy about having to share their family with an intruder. Pigs are herd animals that have a pecking order and only one pig can be the leader. The two pigs must decide who is going to be Top Hog. Once this is established, the pigs can become friends.

The pigs must decide between themselves who will be Number 1. You cannot influence who the alpha pig will be. The pigs may have a very physical scuffle in order to establish this. There may be posturing, head swinging, ear chomping, and nipping until one pig retreats. It may seem very violent but most wounds, if any, will be superficial scratches. Consult your vet if there are any torn ears or puncture wounds.

If the new pig is much smaller in size than Sweet Pea, there may not be much of a battle. The new pig will

wisely accept Sweet Pea as queen and retreat with her pride still intact. If the pigs are equally matched in size or personality type, it may take longer for them to decide who will ultimately be on top. In order to cut down on the passion involved in the competition, there are several things that you can do to make the initial confrontation less intense:

1. Introduce the pigs on neutral territory. Allowing Sweet Pea to meet the new family member on neutral turf will greatly decrease her fervor. She will not be defending her territory, so she will be more open to meeting another porker. If you can take the pigs in kennel crates to a park, put them on leashes, and let them sniff each other, this would be ideal.

2. Put off the initial physical encounter for as long as possible. If you must introduce the pigs in your home, allow the pigs to see and smell each other for a week before putting them together. Using baby gates works very well to separate the pigs. This will soften the initial confrontation. The more the pigs can get used to each other BEFORE they have a chance to fight, the less violent the fight will be. The new pig will seem less threatening to Sweet Pea.

3. Monitor the initial physical confrontation. Use a piece of hand-held plywood or a garbage can lid to separate the pigs if they become entangled. Put the pigs in different rooms if they are too aggressive, and wait a bit for the next confrontation. Go gradually. Make sure that they have a safe space to spar where they won't get injured by slipping or falling. A grassy backyard is ideal.

4. Support the pig that appears to be dominant after the initial confrontations. Scratch and talk to him first when the two pigs enter a room. Feed him first. This works well because it will hasten the ultimate decision of the lesser pig to retreat and make him be content in his position of Number 2.

5. Do not either blatantly reward with food or punish the piggers in order to manipulate the pecking order. Doing so will only postpone the inevitable. Do not feed the pigs together for at least a few weeks. There's no reason to tempt fate and risk instigating an upheaval between them.

6. If the pigs stop fighting at bedtime, allow them to sleep together after the initial confrontation. All competitions are off after sundown! Potbellies have a natural instinct to bed down with partners in order to protect themselves and keep warm. They will even nap with pigs they don't like. This bed sharing allows the pigs to get used to each other. Monitor them again when they wake up in the morning.

7. Do not be surprised if, months or years later, the pig in second place attempts to challenge the Top Hog. The pecking order is not carved in stone. If the Number 1 pig becomes ill or seems vulnerable, he could be challenged to prove that he's worthy of top ranking. If a pig's odor changes from going to the vet or having a bath, this could provoke an attack too. The other pig will react to the strange smell and think that the pig is an intruder in the herd.

8. If you can establish *yourself* as Top Hog, by strong leadership, you should be able to control this occasional flare-up fighting by separating the pigs and initiating timeouts. Any other new pig arrivals will start at the bottom of the pecking order and they must work their way up by posturing.

Jumping Up on Owner's Lap

> "I trained our piglet, Oink, to jump in my lap while I sat on the sofa. It was so cute! Now he weighs 50 pounds and is leaving hoof-shaped bruises on my legs. What should I do?"

Oink is jumping in your lap because you reinforced him for doing so. There is one thing more difficult than training a pig . . . and that is UN-training a pig. However, it certainly can be done.

Do not punish Oink for hopping up. He is just showing affection and obediently responding to past training. Instead, when he gets that certain pose, right BEFORE he makes the springing movement, say, "NO,Oink!" and extend your hand straight out with your palm outward toward Oink. This hand motion will send the message to Oink that you do not want him to jump. Then get up and lead him away. Immediately, walk him back to the floor area where he originally was and give him a soothing tummy rub as you sit on the couch. After a few times of

doing this, Oink should volunteer to roll over for a scratch instead of being led across the room and then back to the couch. The stubby porker may be relieved that he doesn't have to catapult up to the sofa to get some lovin'.

If he still manages to jump up without your catching him in time, place him on the floor, say, "No jumping" and ignore him. Use the hand signal to discourage jumping. If he doesn't jump, say, "Good pig" and praise him.

Be consistent.

In the future, only tolerate and reward the pig for performing a feat if you have given a verbal command. That way you have more discretion. You will be able to control when and if a pig jumps up or not. The pig will not jump up, unless he is asked to.

You may wish to teach Oink not to jump on the sofa, even when you are not on it. Many older pigs have fractured their delicate legs jumping off couches or beds. Simply coax or lift Oink off the couch the minute he gets up on it and say, "NO." Occasionally, give him a treat once he gets to the floor, but only very sporadically. Use tiny treats. If the pig can taste the treat, the size is large enough. You don't want to end up rewarding him for getting on the couch, then quickly jumping off. It would defeat the purpose. The idea is to encourage Oink to stay on the floor by making the floor seem pleasant. You may want to put a soft blanket or pad on the floor for him.

Owners of piglets should remember two rules:

1. Never teach a 10-pound piglet any behavior that would be inappropriate when he tips the scales at 80 pounds.

2. Never wait until the piglet is almost full-grown to teach good manners and obedience. Would you wait until your children are teenagers before disciplining and training them?

Piglets can be trained as soon as they are weaned.

Knocking over Dining Room Chairs

> "My pig knocks over the dining room chairs all the time. Why does she do this? And how can I make her stop?"

Pigs are natural foragers. They love the excitement of looking for something palatable to eat. Pigs are also natural "pushers" and use that beautiful snout as a tool to find hidden or forbidden goodies.

Your porker is probably bored and attracted to the area under the dining room table because there could be some delicious crumbs that have been spilled there. And, of course, it's even MORE fun finding a tiny piece of food under a chair leg. Those pesky chairs get in her way of "vacuuming" the floor too! It's a game for the pig to visit the area, after every meal probably, and see what has been left for her to scavenge.

You must catch your pig in the act in order to modify her behavior. When you see her start to approach the area directly adjacent to the dining room table, startle

her and say, "Leave it!" very sternly. Let her know that this behavior is unacceptable. Gently lead her away to an area that has a rooting box in it, and reward her with a very minimal treat the first time.

A rooting box is a low, plastic or rubber box, filled with round stones or river rock. The owner should sparingly sprinkle a few raisons, birdseed, or popcorn kernels in it every couple of days. Do not put too much food in it. You want to satisfy the pig's urge to push on moveable objects. This rooting box allows the pig to root without being destructive to the yard or house. In the future, repeat this "leave it" command every time you catch her approaching the chairs, but do not reward her every time for going to the rooting box because this could teach the pig to expect to be reinforced for getting into mischief.

Vacuuming under the dining room table after every meal will take the swine's motivation away to knock over the chairs when you are not home. You might want to keep a little hand-operated, battery vacuum next to the table.

Supplying your pigger with a substitute object to nudge on will satisfy her need and allow you to keep your furniture intact.

Lingering Pig Refuses To Walk

"Tusk refuses to walk. I used to take him on nightly outings. I would walk behind him, forcing him to move. Now he whines when I do this and will not stroll at all unless I get behind him. Going on walks is no longer fun for either of us. What do I do?"

Tusk is a typical pig. Only positive reinforcement motivates pigs in the long run if you are attempting to train them. By walking behind Tusk and pushing him forward with your body, you were using negative reinforcement and coercion. Tusk had little choice in the matter. He repeatedly refused to move because it became a challenge for him to make you work for every step that he took. The perturbed swine didn't like being intimidated, and now he is literally digging his hooves in.

Getting a pig to walk can be a challenge if he is overweight or old. The secret is to let it be the pig's idea and not to force the pig. Teach Tusk to both "come" and "follow" before you take him on his jaunt. These two commands will allow you to always enjoy a walk with an enthusiastic oinker.

Teaching "follow" is simple. Take a special treat that Tusk rarely receives. Let him smell it, say "follow," and bait him into walking only a few feet in order to get the prize. Then ask him to walk 3 feet, work up to 6 feet, then 10. "Follow" should only be used for special

occasions. For examples, your pig is afraid to move, it's hot outside, the pig is tired, etc.

If you overuse "follow," it will lose its effect. Save it for special circumstances, and always use a special savory treat that the porker does not normally receive.

When you routinely walk Tusk, use the command "come" in a firm voice. If you usually give him a treat for this, he should be motivated. Let it be his decision. Do not push him or force him into doing it. Pigs love food, and almost any pig would rather come and get the treat rather than just stand there. Pigs won't turn down a free piece of food, unless they feel that they are being bullied or that they might face danger. Let the pig think that it's HIS idea to come to you. When he does, pay him.

Usually the only time that pigs refuse to do simple behaviors is because the owner tries too hard or is impatient. Or the owner is indiscriminately doling out treats to the pig. The pig has no motivation to work for them. Offer Tusk a treat and call him. Pretend that you have a contract with Tusk. Every time he responds to the word "come," he is paid, even if it is just one kernel of popcorn. The pig will understand what is required of him and have the option of walking and being paid or just standing there.

Use the command "follow" for going up steps, difficult pathways, or out-of-the-ordinary situations where Tusk may feel threatened.

When a pig has a choice, usually praise and food win out.

Manners—Piglet Lacks

“My new 5-week-old piglet nudges very hard and nips my hands when I give her the bottle. She wants more milk! She also roots on and bites my arm when I try to ignore her. What can I do?”

You are wise to nip this behavior in the bud. Inappropriate behavior can start at very young ages in potbellies. Even one-week-old piglets can nip.

Now is an excellent time to show the piglet that you are not a littermate to compete for food with, nor are you a milk-vending machine. You are the piglet's “mommy” and leader. You deserve polite respect.

The next time your little piglet nips when you are feeding her, say, “NO, Ouch!” sharply, and use the palm of your hand to tap her snout tip if she persists. The idea is not to hurt or frighten the young piglet. You simply want to express to her your disapproval of her prior actions. Be gentle but firm at the same time. Then take away the bottle for at least half an hour. When she roots on or nips your arm too hard, follow the same procedure.

Baby piglets, if they have been taken away from their moms too early, love to rhythmically nudge on human skin. It duplicates the normal action of a piglet rooting against mom's tummy in order to get her to let down milk for nursing. It can help the bonding between the caretaker and piglet, and is a pleasant, natural experience.

But there is another type of nudging . . . forceful nudging is a way of DEMANDING attention or food. Their tough little noses can actually leave bruises! Learn to tell the difference and take appropriate action.

It may be difficult to think of a five-week-old piglet as capable of being manipulative, but she certainly can be. The good news is that pigs are also very trainable at this relatively young age.

Piglets that are weaned too early (under five weeks of age) do not get the valuable social training from the sow and the littermates that they normally would have had. The piglet that is taken away from her natural porcine family thinks that she is the center of the universe and that everyone is put here on the Earth to fulfill her needs. It will take extra effort and discipline on your part to ensure that the piglet comprehends that she is a pig, not a human in a pig costume. You will have to teach the piglet that there are house rules and boundaries, just as the sow would have done.

Bottle-feeding is not usually the recommended choice for feeding orphaned or early-weaned potbellied pig babies, for several reasons. The piglet starts to think of the human supplying the bottle as a pesky obstacle between him and his ultimate goal . . . drinking from the bottle that you are holding. So the pig does his darndest to show the person holding the bottle that he wants HIS way, and he's going to use aggression to get more food. Also, piglets are prone to aspirating (taking liquid into the lungs) when they nurse off a bottle. They tend to sloppily gulp down the milk, being the charming, greedy critters that they are! This liquid inhaled into the lungs can cause the baby to get life-threatening pneumonia.

Your adorable new piglet can learn to be pan-fed very easily. Just gently dip her little snout into a very shallow pan of warm milk. Repeat this until she does the drinking herself. It may take a few tries.

Enjoy your baby! There is nothing in this world more appealing than a newborn, potbellied pig.

Nipping Fingers When Eating

"My 6-month-old darling, Spam, nips my fingers when I feed him. He never used to do this! What's wrong with him?"

There are several types of nipping behaviors seen in potbellied pigs. Much of how pet pigs behave relates to how they are treated by their caretakers. They take their cues from their owners, and it shapes their future behavior patterns. This is especially true when owners feed them.

When pigs come to their new homes as tiny piglets, they are very hesitant to approach their new owner's outstretched hand, even if it has food in it. The piglet wants the food, but he doesn't trust his caretaker enough to risk being grabbed. As the pig learns to trust his owner, he will gingerly take the food and display good manners, at first.

What comes next is contingent upon how the caretaker handles feeding by hand. If he doles out treats on command, one after another, the pig will look upon him

as a slot machine that rewards him with a 100 percent payoff. Pull the handle and it's yours!

So the porker, being the grocery-loving beast that he is, pulls faster and harder on the handle that dispenses the free yummies. This results in bitten fingers and a lunging, rude piggy. He not only expects to be fed, he now DEMANDS it. Pigs will often use this form of manipulation with very young children, who, by the way, should never be allowed to hand-feed a pig.

The human often gets a little intimidated by the nipping, so he dispenses the food faster. The pig learns to snap even faster. So the human gets angry and yells at the pig or throws some of the food on the floor for the porkster. Now it's *really* getting fun for the pig. Not only is he getting tons of goodies, fast and furious, but he now has an exciting game going to see how fast he can nip the hand that feeds him.

The pig is in control.

A pig should never get a treat without earning it. If he is taught this very early, he will not expect a free lunch, so there is no reason to nip a person's hand. As a pig is given a goodie, the owner should say "treat," signifying to the pig that he has been paid. Clapping your hands together is even better.

This tells the pig that there's no percentage in him pestering or begging for more. It acts as a release. The plug on the slot machine has been pulled, for now.

Some young, over-anxious, excited pigs will mistakenly nip your finger right as you are handing them a treat. This should have been corrected the very first time that it happened. The recipient should immediately say, "Ouch!" and then, "NO!"

No more treats should be dispensed for at least an hour. Most pigs learn very quickly not to hurt a person's hand, by being careful, when taking treats. Use larger pieces of food while training the pig to be more sensitive.

When hand-feeding a pig, never suddenly pull back your hand or tease the pig. Doing so will train your pig to lunge and bite whatever he can get hold of. It's not fair to the porker, and it's bound to backfire.

Be consistent when handing out food, and use verbal signals to call him over, ask him to earn the treat, and tell him when he's been paid. This will teach your pig that you are in control of dispensing food and that he will only get the jackpot if he shows proper manners.

Nipping When Cranky

"Runt seems to be nipping more and more. Sometimes, when I put his harness on or when I disturb him as he makes his bed, he bites. He also bites when I give him a bath. What can I do?"

There are many reasons why pigs and other animals bite. By far, most potbellies do not bite and are docile, gentle animals, especially when compared to dogs. But some housepigs occasionally DO nip, as with virtually all companion pets.

1. Pigs can bite if they are prone to territorialism, especially if they are spoiled or live indoors as "only pigs." This can be manifested by charging and intimidating houseguests or family members. (Please refer to Chapters 10 and 11 on aggression.)

2. Underfed pigs can act bitesy, for obvious reasons. Pigs also can nip in order to manipulate the human into dropping the food or giving it to them faster. (See "Nipping Fingers When Eating.") Pigs can accidentally nip their caretaker's fingers when they are excited about being hand-fed.

3. Pigs can bite to defend themselves and their young or to establish their herd status among other pigs. Frightened or abused pigs may bite also.

4. Pigs can bite as a warning to leave them alone because they are not feeling well or are in pain. Therefore, it is a good idea to take a pig that is biting to a vet in order to eliminate any possible physical causes.

5. Pigs that are weaned too early (prior to 5 weeks) can nip. These piglets, however, are NOT necessarily predestined to be bad pets. With proper discipline and training, they have the potential to be good companions.

6. Pigs can snap because they smell the scent of another pig on your hand, are jealous, and want to get your attention. This can occur when an owner brings a brand new pig into the house and the older pig resents it.

7. Pigs can bite because they have learned to get their way by biting. It is a method of crudely communicating to a human being what they do or do not desire.

8. Pigs can nip out of curiosity. They love to sniff shoes, people's feet and fingers. Once in a while, they will indulge in a gentle exploratory nip as a "test."

Runt's nipping falls into category number 7. Your pig is using biting to communicate to you that he'd rather not have a halter on, he'd rather not be disturbed as he is settling down for his night's rest, and he certainly doesn't enjoy the thought of a bath. Communication is desirable between owner and pet, but this method of forceful communication cannot be tolerated.

The next time Runt nips you, firmly say, "NO. Bad pig." Tap his nose stopper with the palm of your hand. Wait one or two minutes for him to calm down, then remove him from the room for a timeout. The timeout will allow him to settle down, but it will not cure the problem. You must consistently correct Runt every time he nips, and then take him from the room. Do not be intimidated by his nipping or give in to it.

When Runt is having his bath and is not hostile, praise him and show him lots of affection and caring. Do the same with the other provocative situations. Spoiling Runt will encourage him to communicate by nipping because he will be used to getting his way, then he will start to demand it by biting.

Let Runt know that you love him but that *you* are setting the rules.

Opening the Refrigerator

> "Rosie has learned to open the refrigerator! I never dreamed that she would be able to figure it out. Our dogs sure can't. Now what do we do?"

Pigs have very good noses and love food, obviously. The result is that not many refrigerators in pig abodes go unscathed! When it comes to groceries and pigs, sometimes it's more effective to remove the potential of having a problem rather than attempt to fight the pig's lust for eatables.

A double-door refrigerator is virtually pig-proof. There's not room for that tool-like snout to wedge between the doors that come together. A single-door refrigerator is vulnerable. The porkster can simply push the door open at the side of the refrigerator where the door meets the main structure.

Purchase a piece of solid, strong Plexiglas at a hardware store. A piece about 2' x 1' will do the job for even tall piggers. Also purchase strong adhesive glue that is suitable for plastics. Adhere the piece of plastic to the SIDE of the refrigerator so that it is extended flush to the front of the refrigerator door. The pigger should not be able to get his snout in the small space that is between the door and fridge . . . it is covered with the Plexiglas! And yet the door should open easily for family members.

It is probably a good idea to feed new pet pigs

somewhere else besides the kitchen. Even when you are cooking, it is not the best idea to allow the pig to snack on recipe ingredients. If a pig is never fed in the kitchen, he will not be obsessed with opening cupboards, pantries, and the refrigerator all the time. You will be able to cook without being pestered by an overly-demanding porker, and the pig will be taught that food must be earned, not stolen.

Piglet—Out of Control

"Our sweet, new, little piglet is no longer controllable. She used to be timid, but now she runs all over the house causing trouble. She gets into everything and is making the house a mess. How can we "tame" her?"

A new piglet should be taught the rules of the house on a very gradual basis. Her world should be expanded in a controlled manner as she learns manners and boundaries. It sounds like your pig was given too much, too soon. The discipline was too little, too late. The good news is that she can still be adequately trained if you back up a bit.

Start confining the pig to one room of the house in order for her to learn some respect for it. Baby gates can be used to separate the rooms.

Ordinarily, when a new piglet is no longer frightened

and has learned to trust her new owners, she should be introduced to the various rooms of the house that she will be allowed in, one room at a time. Retrain your piglet by starting the process over.

When inquisitive piglets first begin to explore the house, they tend to knock over or displace household objects and chew on things. This is the time to start teaching the pig boundaries. Pigs are curious critters, and they need something to do when confined to the home.

It is the owner's job to guide them to the right area and discourage them from off-limit areas. At the very first start of any inappropriate behavior, the pig should be disciplined. Pigs are creatures of habit. Once they start a bad habit, such as opening cabinets or chewing on sofas, it is more difficult to change them.

At first, it might appear to be "cute" when a pig attempts to get in the refrigerator or spin the Lazy Susan with her snout, disrupting cans of vegetables and pickle jars! However, this mischievous behavior is bound to escalate unless the pig is immediately diverted from it and taken to a pig-proof room, a rooting box, outside, or a place where she can't be destructive. She should be verbally reprimanded with a low, controlled voice as you stand tall and maintain eye contact.

Say, "LEAVE IT," and lead her away. This is supposed to be a learning experience, not a punishment at this point. Piglets are very sensitive, and the owner shouldn't frighten them or make them recoil. Be careful that the only time you talk to your pig is not just to reprimand her. Pigs need praise as well as discipline. These clever, perceptive animals are able to understand a lot of what we say.

Rewarding the piglet's acceptable behavior with belly rubs, tasty morsels, and plenty of loving attention will go far. Constantly screaming at the pig, not being consistent, and not having strict boundaries will confuse the new arrival. Piglets need firmness, but they also need kindness and patience too. It is a delicate balance, and the most successful caretakers are very in-tune to their pig's reactions and what those reactions mean.

Potbellied piglets love routine, structure, and simplicity. This is good news because structure can be used to establish good potty habits as well as appropriate behavior in the house. Try to feed the piglet at the same time every day, let her outside at the same time every day, and structure training sessions at the same approximate times, day in and day out. Piglets do best in households where surprises and chaos are kept to a minimum. Small children seem to bring out the worst in pigs. Piglets don't like to be handled in the same manner as puppies do. They need gentleness and a place to go where they can be left alone. They like their bodies to be respected and to be in control of the physical handling. Roughhousing, probing, or chasing is not appropriate for the sensitive pigmies. Piglets take nothing for granted and will only trust you after you have proven yourself.

Training a new piglet to sit and spin (see Chapters 7, 8, and 9 on training) will establish you as her leader and help the piglet become confident.

It will also help with bonding and proper house manners. Pigs need to use their minds because they are the most intelligent of all domesticated animals. If they receive no mental stimulation through training, they will improvise and find another outlet themselves, such as

the challenge of tearing a piece of carpet up or removing a piece of wood from a cabinet.

Piglets have a lot of energy. Teaching a piglet to "come" is a must, and asking her to run across the yard for a special treat will even prepare her for "Hog Calling" contests at pig shows. The more time that you can spend with your piglet, the easier it will be to help shape her personality into a docile, content, house-friendly pet.

Potbellies need time outdoors, and they don't usually do very well living in apartments and condos. Pigs don't need tons of space, but they do need a backyard where they can just be a pig, snorffling along the ground, rooting, smelling intriguing new scents, and grazing on the fresh grass. This satisfies some of their basic needs that otherwise might make them aggressive or unruly if left unfulfilled.

Taking the pig for a walk to the park or to the local school grounds will enhance her world too. The more non-destructive activities you can encourage, the less time your pig has for destructive habits, such as chewing household items. Having an indoor-outdoor living arrangement is often very successful.

Leash and harness training will help civilize an unruly pig and enable you to take her places and expand her world. Making her earn her treats will result in her giving you respect and listening to you.

You have allowed your piglet to run willy-nilly throughout the house, doing whatever she wishes. By retraining her and starting over as if she had just arrived, you will be able to impart to her that the house and your belongings are to be respected. She will have new boundaries, plenty of mind-stimulating training and exercise, lots of understanding and love, and she will get a new sense of security from being put on a routine.

Quelling Frightened Piglets

"My two new piglets are frightened to death of me! How can I make them like me and be good pets?"

A good breeder will socialize the pigs before they are sold so that the piglets are used to being handled. A prospective buyer should always insist upon this. It cuts down on the stress that the piglets must endure and allows the bonding between the caretaker and piglet to begin immediately.

Baby piglets are born with a strong sense of self-preservation because they are vulnerable to being snatched up and eaten by predators. Your piglets are hesitant to approach you because they don't know you are their friend. Also, they have no reason to approach you. They'd rather play it safe, run from you, and hide.

The piglets need to trust their caretaker and must have a desire to approach and spend time with the caretaker. In time, if you are kind and just sit back and wait, the pigs will probably come to trust you. But there are steps that can be taken in order to speed up the process. The same steps can be taken with frightened, rescued, or abused pigs.

1. For best results, separate the piglets so each one lives alone while you are working with them. The piglets must be kept in separate playpens or rooms in order to speed up the trust-earning process. They have little motivation to get to know you and have no need to depend upon you as long as they have each other. They need to be encouraged to act as individuals instead of an apprehensive little herd. As long as they are together, they will not pay particular attention to the affection and dependability you are trying to convey.

2. Place your hand around each pig's bowl, while they are separated, in order to motivate the pigs to approach you. Take advantage of the porker's lust for eating, and use it as a tool to get the pigs to come to you. Hold their food bowls in your hand, and patiently wait for the pigs to move forward. If they don't approach, take the food away and come back in 30 minutes.

It may take a few sessions before they will sneak up to the bowl. Extend your arm out as far as you can, then softly say "come" and the piggers' names. Make yourself small and less threatening by crouching over and staying low. Tell them how adorable they are and that you won't hurt them. Let the pigs sniff your hand and the bowl. Whisper to them. Quietly, let the pigs eat and make no advances toward them at this time. Keep this up for one or two days until you see the piglets displaying more confidence. Feed the pigs at the same time every day. Keeping them on a routine makes them more secure. They know what to expect.

3. Start offering the pigs treats from your hand and continue to say "come" so they know what to expect. After a few sessions of this, carefully and slowly try to scratch the pigs' chins as they munch a goody from your hand. Keep your arm and hand as low to the ground as possible. If they allow this, praise them and give each of them a treat. If they retreat, save the handling for another session. Always let the pigs come to you. They need to feel as if they have some control.

4. Be very gentle and predictable around the new piglets. Don't let other household members frighten or startle them. Speak in your mildest voice, and don't make quick gestures. Piglets are very motion-sensitive. This is because they are always leery of possible sneak attacks from enemies that loom above their heads, ready to grab them. Always be squatted down, so you look less formidable.

5. Hold out your finger or hand while saying "touch." If the pig sniffs or touches it, softly praise and reward him. If he retreats, try again later. Be patient.

6. Start scratching the pigs' undersides very gently with one finger. This will put most pigs in ecstasy. However, they may not roll over for a belly rub just yet. A pig likes to stay on his feet when danger could be present so he can run away if necessary.

7. Hold off on any bathing or picking up the piglets because it will set the socialization process back. Let the pig learn to trust you more before you begin these

unfamiliar and stressful actions. First, make sure that he knows his name, the command "come," and that he will eat from your hand.

8. Whenever the piglets approach you on their own, reward them with a few feed pellets until they are fully socialized. Bond with them by spending lots of time stroking or massaging them.

9. Be sensitive to their feelings. If they are retreating, you are moving too fast. Never corner or grab the pigs, and try not to make them run. Being consistent in your actions will make the pigs trust you. And the food will speed up the process in getting these shy, vulnerable animals to bond with you.

Refuses To Go Out in Bad Weather

"Hoofy refuses to go out in cold or snowy weather for his potty break. I've tried pushing him and now he's even worse. How do I get him out? He weighs 150 pounds!"

Vietnamese potbellied pigs lived in the tropical jungle originally. They prefer temperatures of 60-70 degrees. Asking a content, warm pig to get up and waddle outside, naked, in the cold, wet snow will not be met with much enthusiasm from the complacent porker.

Start with a positive attitude and try to convey this to the pig. Because it is natural for a pig to not want to go out in the cold, he deserves some extra treats in order to motivate him now and in the future. Make him remember the experience as enjoyable, not an annoying power struggle. Get a coat made for Hoofy that will keep him toasty.

Pigs are stubborn animals that need positive reinforcement. If you try to force him out, he will probably dig his hooves in, remember the experience as a miserably negative one, and fight you more the next time.

So instead, give him one of his very favorite treats for going out (like a very small piece of cheese, a tiny sliver of pizza, or a cookie). But do not bait the pig with it, except for the first time. Lead Hoofy out the door with it to the proper place the first time. In the future, Hoofy only gets the cookie if he goes ALL THE WAY out the door and to wherever he is supposed to go. This is very important. If you continue dispensing treats to him for getting to the door, he has no motivation to actually go OUT the door.

When Hoofy reaches his destination, get excited, say "good pig," and do some happy talk, pat him on the head, and scratch him. Try to impart to him that he has done something very special! Give him the rest of the cookie.

From then on, give him the same special treat if the weather is bad and he goes outside. Either give it to him yourself or hide it in the snow beforehand in the general area outside. Then it will be a game to him! The idea is to make the pig think that it's HIS idea to go outside. Make Hoofy identify bad weather with his earning a very succulent reward that he rarely gets.

This works because pigs will tolerate doing difficult things if they get plenty of praise and a special reward. When pigs do things that are uncomfortable or against the nature of a pig, it is not spoiling them to "up the ante."

Roommate Problems

"Toby doesn't seem to want to have anything to do with my roommate. My roommate approaches Toby, and he snubs her and just walks away. Toby centers all of his affection toward me, and my roommate really wants to be his friend. What do we do?"

Toby has no motivation to befriend your roommate. It's time to change that! The next time your roommate and Toby are in the same room together, have your roommate slip Toby a very tasty treat that he never gets — a very small bit of cheddar cheese, for example. (Treats that are rich in calories can be given sparingly for special purposes if the porker is not obese.)

At least once a day, have your roomy give Toby the same prized goodie. No one else should give Toby this special treat. In no time at all, Toby should be approaching your roommate. Have her praise Toby and eventually ask him to earn the treat by doing a simple trick, such as spinning in a circle. Encourage your roommate to learn to communicate with Toby by talking to him and handling him. This works well because pigs sometimes need to be strongly motivated to expand their horizons and, of course, scrumptious food works very well for this. Some pigs like to play it safe and spend all their time with a familiar face. But once a new person introduces food and games, the pig is happy to gravitate toward him or her. This allows the pig to get to know and trust him or her.

Rooting the Lawn Up

"Wilbur is destroying our lawn! Should I put a ring in his nose? What can I do to make him stop doing this?"

Pigs are natural rototillers. This is how they obtain their food in the wild and spend their days. They are also curious critters that want to know what is UNDER everything! Ringing potbellied pigs' noses may not be the most humane way to solve the problem. The ring will cause discomfort and pain to the porker. He is only doing what is a natural behavior for a pig. It will also be unsightly and is not allowed at sanctioned potbellied pig shows.

Any type of trick training, especially ball games (such as soccer), will teach Wilbur to like more "sophisticated rooting."

Putting chicken wire or old chain-link fencing flat against the soil under your lawn and fastening it down securely, works beautifully. The grass will grow over the fencing, and you will not even see the wires. The fencing

will prevent Wilbur's snout from turning over the soil, but it will allow him to graze and forage a bit.

Making a portion of the backyard, nearest the house, spread with pea gravel will cut down on turf rooting. You can sprinkle popcorn or birdseed over the gravel, giving Wilbur a challenge to find each morsel. It may take him several hours, at the very least, before he decides to move on to the turf.

Some pigs are just grazers, not rooters. Taking good care of your lawn by adequately and safely fertilizing and watering it will make it grow better. The pigs often seem satisfied to graze instead of rooting if the grass is long, green, and lush.

Or you may want to designate one special corner of the lawn where it's okay for Wilbur to root up a storm. You can re-seed it and allow him to go into a new area.

Pigs really do need to do some tricks or rooting; otherwise, they need a substitute activity to enrich their environment.

Rude Female Pig

“We have a 4-year-old, intact, female potbelly, Bertha, who is obnoxious when she comes into heat. She jumps up on us while she is standing on her hind legs. Our vet says that she's too old to spay safely. What can we do?”

Female potbellied pigs come into estrus about every 21 days. They can exhibit inappropriate behavior for as long as three days to an entire week, per month, as the hormones are building up and winding down. In order to make good pets, all potbellied pigs should be neutered or spayed. The older females can present a spaying problem if they are not operated on at less than one year of age. The surgical procedure is more risky and expensive as the pigs grow older and put on unnecessary weight.

However, just because intact females are motivated by hormones during part of the month doesn't mean that poor behavior should be tolerated. Females can be taught to go through the entire month showing good manners!

When Bertha jumps on you, take hold of her front legs and just stand there holding them in your hands. After a minute or two has passed, the passionate lady will decide that she doesn't like being restrained and held up and being in someone else's control. She will struggle. Don't give in for a few minutes and then gently let her down.

She may try to jump on you a few more times. When she does, say, "NO. Bad pig." Repeat the training exercise. This works very well. If you are sitting down on a couch and she jumps the couch, repeat the exercise.

This works because the pig's distaste for being restrained is more powerful than her desire to jump on humans. Ideally, all of the family should join in and teach the pig that she must show respect toward humans, even during heat cycles.

Slippery Floors

"Burp, our new piglet, is scared to death of the hardwood floors in our house and won't walk across them. I can't get him outside to go potty. I have to carry him. He's getting heavy, and my back is getting sorer every day. What should I do?"

Pigs, especially piglets, are frightened of slippery surfaces, like your hardwood floors, because it gives them poor traction. They could fall and be injured. Their little, hoofed feet invite traction problems!

Also, potbellies like to have their hooves on a solid surface, in case of danger. If they need to run (their only defense), they can peel out!

Burp is very young and unfamiliar with slippery surfaces and the proper adjustments that he must make in order to walk on them safely.

He will need to be counter-conditioned by you in order to get the knack of keeping his balance on the wood floors. As with all Pig Programming training, it's best to break the desired behavior down into the smallest possible increments.

Burp needs to be desensitized to slippery surfaces gradually in baby steps.

First, start by taking all the throw rugs in the house and covering the wood floor with them, making a path where Burp can easily walk across it and not slip. Get down on your hands and knees and with a super treat, like cheddar cheese, gently attempt to coax Burp across the carpeted path. Give him a lot of treats and praise when he reaches the other end! This will motivate him to want to go all the way across instead of stopping in the middle or turning around halfway.

This may take several attempts. It is very important that Burp is not forced and that he feels safe. Burp should be able to trust you. Let it be HIS decision to cross. Don't push him.

Once Burp has learned to cross, take away one of the rugs, which will make a bare spot. Fill it in and use the extra space to make small, even spaces between each rug. There should be some space between each of them now.

Coax Burp into crossing. He will probably be a little leery of the exposed surfaces and will scurry between the throw rugs. Reward him for each rug that he crosses.

Make sure that he is confident in crossing before you remove the next rug. Gradually, widen the exposed floor surfaces. The idea is to let Burp get some physical expertise and confidence walking on the new surfaces. He will learn not to slip. If he refuses to cross, back up and bring out a carpet. You are going too fast.

Soon you will be down to only one carpet. Reward Burp only at the "finish line." After he has mastered that for a few days, remove the last carpet.

Traveling by Plane

" What can I do so that my pig will not be frightened or uncomfortable during air travel? **"**

Preparing the porker for the trip ahead of time is an excellent idea so that the pigger will not be stressed.

1. Purchase a large, airline-approved kennel that is sold at pet stores. It should be roomy enough to allow the pig to walk a few steps and turn around in without hunching over or struggling. You need to crate-train your pig by taking the kennel apart and feeding her in it. Place a bowl in it with a trail of snacks leading to it. Let her discover it on her own. Allow her to get used to being in it, and keep feeding her in it with the top off. Let it always be her choice to enter or leave the kennel, and make it a positive experience for her.

Pigs tend to feel trapped in unfamiliar, confined spaces. If she is allowed to be the one in control of the situation, she will learn to love the crate instead of fearing it. Be patient. Put her favorite bedding in it as well as surprise snacks. Let it be her sanctuary from the world.

2. Get her used to your closing the door to the crate while she's sleeping in it. Let her out if she awakens and wants out. Gradually, get her accustomed to having the door closed and then start slowly moving and sliding the

crate, very gently. Have someone help you pick it up if she seems secure and comfortable. She should start actually enjoying her new "nest" in the crate. She should need no tranquilizers for the trip. Pigs tend to fare better on airplanes than dogs, who are bothered by the loud noises. Pigs tend to sleep through the whole ordeal!

3. Call your vet and make an appointment to get a health certificate and blood testing if you are going out of state. Most states require this because of the spread of possible porcine disease across state lines. Be sure to carry these documents with you the day of the flight. Your pig will fly in a pressure- and climate-controlled baggage section of the airplane. Ask your vet if the pig should have food or water withheld before the flight.

4. Be sure to arrive at the airport early on the day of departure. Checking in the pig will cause a lot of attention! Using plastic wire wraps to keep the porker's crate door shut is a good idea. Towels or straw both make excellent bedding because they are so absorbent. Put enough bedding in so the piggy won't be injured if he is suddenly jostled in his crate.

Flying pigs in hot or very cold weather is not a good idea, especially if it is a connecting flight. The airlines must follow USDA temperature restrictions for handling animals on the ground. No animal can be exposed to temperatures of less than 45 degrees or more than 85 degrees for more than 45 minutes. Be sure to check the weather report before your trip. It can get very warm on the tarmac, and pigs are vulnerable to heat stress. A direct flight is always best.

Bring extra bedding, liquid cleanser, and paper towels for use when you clean out the carrier after the trip is over. You will be required to fill out and attach adhesive forms to the crate indicating if you want the piggy fed and watered by airline personnel. Be sure to include your telephone number and address. It's prudent to attach little plastic water and food cups to the inside of the crate door so the door doesn't have to be opened for them to service your pet. Otherwise, the pig could easily get frightened, jump out, and hide in the warehouse.

5. ALWAYS go to the stewardess when you board, and have her check with the captain to see if your pigs are actually on board. By law, the captain is supposed to be aware of how many animals are traveling in the hold. This will remind him that animals are traveling on the flight, and it will also allow you to be sure that the pig is not "lost" somewhere. The special baggage hold for animals will be kept at the same temperature and pressure as the cabin of the airplane.

6. Be prepared to wait at least 15 minutes after the plane has landed to pick up your pig at the other end. Bring some special treats for the little porker. He will be very glad to see you!

Urinating on Deck

> "Hambone is urinating on the deck. He used to walk down the deck stairs and go in the proper spot. Why does he do this?"

First of all, Hambone needs to be checked out by a vet. Anytime a pig changes his urination habits, it is suspect. Urinary obstructions and infections are common in Oriental pigs. Hambone might have arthritis or another malady that makes it painful to walk and climb stairs. If he has become obese, that may influence his walking habits too.

Once a physical reason has been eliminated, we can work on Hambone's habits. Hambone is probably urinating on the deck because he isn't being taken out enough. He can't wait to use the steps or he does not want to go to the trouble of walking all the way down the steps to the normal area. Pigs have short legs and are not athletic. A level exit from the house would be more desirable. However, Hambone can certainly be motivated to go up and down the steps if he has no physical problems.

Start taking Hambone outside more often. In bad weather, pigs will often hold it, waiting until the last moment to waddle out. See if it makes a difference if you take him out every couple of hours.

Otherwise, Hambone has to be taught that he must go all the way down the stairs, and nothing short of that

is acceptable. Start by firmly and quickly leading the pig all the way down the stairs, for a few days, on halter and leash. Say, "All the way down, Hambone." Give him a reward at the bottom, such as a few kernels of popcorn. He needs to be retrained and motivated to go the entire nine yards.

Then, let him out on the deck on his own as you look out the window. The second he stops moving and strikes "the pose," run out and briskly lead him all the way down again while verbalizing the proper command. Repeat this over several days, paying him every time. The pig will learn that you will not tolerate his stopping on the deck. It is very important that his potty breaks be monitored until you are satisfied that the pig has learned to go down the steps on his own. One hundred percent consistency is required. Leaving an occasional "surprise treat" at the bottom once a week will motivate Hambone to go all the way to the bottom.

Pushing or punishing Hambone if he makes a mistake will set the training back. Using positive reinforcement and letting him know that you won't compromise should be successful.

Veterinarian—Fear Of

"My pig, Sassy, hates to go to the veterinarian for her annual hoof-trimming appointment. The vet has to wrestle with her in order to get the anesthesia mask on her. We dread the whole experience. I feel so sorry for Sassy. What can I do?"

The trip to the vet should be a pleasant experience for everyone, and it certainly can be. The first step is to take Sassy back to the vet just for a fun visit if they can schedule you. Let the people waiting in the office know how wonderful pigs are, and let them scratch Sassy and talk to her. Ask your vet if she can come into the exam room and have a few bits of cheese (a strong incentive) and just receive a belly rub from him for a few minutes. Make sure that she gets plenty of yummies and praise. Walk her around the office and exam room, allowing her to associate the smells, sights, and sounds with a GOOD experience. Then take her home.

Go back again on another day before your next appointment, and allow her to again enjoy a non-stressful, positive experience.

At home, practice with Sassy having the anesthesia mask (cone) placed over her face so she can be anesthetized properly. Use a coffee filter or a cup from a fast-food restaurant. Hold it over her snout, say "good pig," remove the cup, and pay her while saying "treat." Gradually, increase the time that the mask is over her nose. This will desensitize her to having a strange object

placed over her snout, and she will associate the experience with something positive (food).

When you do have your real vet appointment, try not to be apprehensive or distress Sassy. If you are upset, she will be upset. Sensitive pigs are often able to read our emotions better than other humans are!

Taking Sassy to the vet in a crate (kennel) that she is used to, with her own special blankets, will help. Otherwise, taking her in on a very hot or very cold day may make for an upsetting, stressful walk across the parking lot. If you can carry the crate in, or have a dolly to roll the crate in with it's a good idea.

Having a vet that is experienced in potbellies is a must. Potbellies get upset very easily, and this can be of great medical concern. Potbellies need a vet who understands the anesthesia procedures and the special risks that potbellies have with injected anesthetics. An experienced vet realizes that potbellies take special handling and are not like dogs and cats.

If your vet will allow it, come back into the room where your pig will be anesthetized in order to calm her. Try not to interfere with the vet's work, but give Sassy her command, "Mask," when it's time to give her the gas. If this works like it should, I'm sure that your vet will be very grateful for the help.

Wakes Me in the Morning—Screaming

> "My pig seems to be waking us up earlier and earlier every day. Now she starts making a ruckus at 4:00 in the morning! We try getting up and feeding her and then going back to bed. A few hours later, she throws a tantrum and wakes us up again. My husband is threatening to get rid of her. Help!"

Your clever little darlin' has you trained, using the conditioned response method. She has learned that only people that are AWAKE have the ability to feed her. Therefore, it is her duty to see that you get no sleep!

You are conditioned to respond to her screaming by bringing out the feed bag. You desire peace and quiet, so you feed her even more. Her antics were successful. She wants more, so she screams again, and it works.

The pig is in control. She has you over a "pork barrel."

It's time to break the cycle and take back the power from the greedy pet porker.

Starting tomorrow, you will decide ahead of time when the pig is fed. You will no longer take cues from the pigger. She will scream her lungs out the first couple of mornings, demanding that you fill her needs. It will be loud and in order to have some peace, you will be tempted to give in.

Don't!

Think of it as an investment in the future. By enduring her screams now, the piggy will learn that screaming doesn't work, and she will stop after a few

days. Set up a rigid schedule of when the pig is fed, the same time every single day. This chow time should be at least an hour after you wake up. This will take away the motivation of the pig to arouse you from sleep in order to be served a meal.

All members of the family must agree on this. If someone gives in after hearing the initial, irritating screaming fit, the game plan won't fly. Every member of the family must be in agreement or you will be training the pig to be PERSISTENT in her screaming.

After your pig is on her new schedule, both of you will get a lot more rest.

Wakes Me Up at Night

"About 3 months ago, Harley started coming into our bedroom and waking us up about once a week. I thought he was having nightmares, so I gave him belly rubs and warm milk. Then he started coming in almost every night, whining. Now he comes in and when I comfort him and put him back to sleep, he wakes up an hour later and disturbs us again. We can't get any sleep! I've tried ignoring him, pretending to be asleep. Then he screams! We aren't zoned to have potbellies, so we don't want the neighbors to hear him squealing. Help!"

Oh, no one has ever said that potbellies are dumb! This clever, manipulative little guy has you at his beck and call.

Actually there are two problems here: one is the night-owl pig, and the other is the possibility of the neighbors hearing the porker screaming.

Harley has to be taught that coming into your bedroom and waking you will not benefit him in any way. He will not get food or positive attention for it.

Instead, he will learn that it can actually inconvenience him!

The next time that Harley awakens you, take him outside into the cold night air and walk him. I'm sure that Harley would rather be snoozing away in the comfort of his warm bed.

Repeat this action every time the pigger disturbs your sleep. YOU MUST BE CONSISTENT. Harley will soon be VERY soundly sleeping through the night!

Many people do not realize that potbellies are illegal in many cities because under the zoning laws they are technically classified as "swine." At the turn of the century, many cities passed laws that prohibited farm pigs from being raised within city limits.

The people that passed these laws had no inkling that someday a mini-version of the noble pig would actually reside in people's abodes.

Some of these laws have been challenged and changed and specifically address the potbelly. However, in some cities, the pet pigs are still illegal.

It is always a good idea to find out first from the zoning department of your city if, indeed, the diminutive cousin of the farm pig is prohibited. Otherwise, a single complaint from an irate neighbor could result in people losing their beloved pet piggy.

Be sure to get a copy of your local zoning laws IN WRITING.

X Marks the Spot—Defecating in the Car

"Fanny must have pottied at least 5 times on the way to and from the vet's office! Our car was a mess. Now we are taking her on a road trip with us, and we don't want to stop every 30 miles to walk her. Why does she do this? She lives inside our home and is housebroken."

For Fanny, going out in the real world, in a moving automobile, is very exciting! Imagine how this expands her world and how amazing it is to her to be in a car as it speeds along, vibrating and making strange noises.

It's the excitement that causes Fanny's bowels to move too quickly. Even if you walk Fanny first, I suspect that she will soil the car numerous times on her next automobile jaunt. To her, it's more stimulating than a roller coaster and she just can't control herself.

In time, Fanny will get very bored with traveling and will revert back to her normal potty habits if she rides in the car often. Eventually, she probably will sleep most of the trip.

Before you leave on vacation, as a training exercise, put Fanny in the car in a kennel and take her to the nearest gas station or grocery store. Clean out any droppings. Turn around and come right home. After three or four test runs, you should see a marked improvement in Fanny's habits.

On the trip, putting Fanny in a large travel kennel,

filled with absorbent, clean towels, will ensure that she is not thrown around the car, which could lead to injury. It will also provide a safe nest for Fanny and give her some security. Rooting around in the towels will calm her down. Since it will suffice as her own comfy little bedroom, she will not want to spoil it by defecating in it. Pigs do not relieve themselves in their sleeping or eating areas, normally. Give Fanny some treats in her crate, and it will further ensure that she respects her new traveling digs by not dirtying them. She will be reluctant to potty in the area where she was just fed.

It's important to train Fanny to happily be able to go in and out of the crate herself, before the vacation. Make sure that the crate is large enough for her to comfortably turn around in.

If Fanny does make a mistake, clean out the dirty towels immediately and replace them with clean ones. Doing so will tell Fanny that the crate is not the proper place to defecate.

When you do leave on your vacation, plan to stop overly often for potty breaks the first day so that Fanny does not learn bad habits. Rest stops along the freeway provide excellent piggy walking and exercise areas.

X Marks the Spot— Defecating in the Swimming Pool

> "Daisy urinates and defecates in her swimming pool even after she's just gone potty outside the pool. I'm afraid that it will harm or kill her because she drinks the water. How can I make her stop using the pool as a big toilet?"

Ah, there's no accounting for some pigs' tastes!

Drinking the water will probably not harm your pig. To discourage your swiner from using her wading pool as a toilet, put a few grapes or carrots in the pool. Because pigs instinctively will not potty near or in their dining area, Daisy will probably not foul the pool due to the fact that it contains edibles.

And she will have a blast bobbing and diving for veggies.

Young Piglet Hates To Be Held

> "Poppy hates to be picked up. But I've heard that it's wise to keep on holding her until she stops screaming or else she will think that she's getting her way if I put her down."

Negative reinforcement rarely works on pigs when you are training them to enjoy a behavior (being held).

It's a natural reaction for Poppy to scream when she is being held. Pigs have a rational fear of being restrained.

Instead of making her tax your eardrums, try positive reinforcement.

Every time you hold Poppy, give her a small treat and scratch her ears or belly. Talk calmly to her or sing a soft song. When she protests, quickly put her on the ground. She will soon learn that she is missing out. Remain sitting there on the floor for a minute, allowing her to either accept or reject being held again. Let it be her choice. If she refuses, come back in an hour with more rewards and hold her. In a few days, she will be jumping in your lap! She will think that it was originally HER idea to be held!

Lifting a pig while she's screaming and continuing to restrain her, is not a good idea because it distresses the piglet and can make her fearful and distrustful.

Rewarding the pig for good behavior will be more effective than punishing her for natural behavior.

Zoning

> "I've heard that potbellied pigs are not legal in some cities. Is this because of their behavior or what? We want to buy one!"

It is always prudent to check out the city or county zoning laws before purchasing a pet mini-pig. Be sure to get the specific zoning laws that apply to your parcel of land, in writing, as proof.

The reason that potbellied pigs could be illegal is that many years ago the city governments outlawed "swine" or "livestock" within the city limits, for obvious reasons. Animals that are used for food or fiber do not belong in cities; they belong in rural areas on farmland.

The people who wrote these laws never envisioned that a "lowly" pig actually could be a good housepet! Up until the mid-1980s, there were no miniature potbellied pigs in the United States. There was no reason to change the laws until these perky, porcine wonders became a fad. Many cities have revised their swine laws and now welcome the Oriental pet-piggy imports. However, some cities still feel that "a pig is a pig," even if it is clean, well groomed, tractable, and housetrained.

Potbellied pigs generally make good neighbors. They don't bark (like dogs), smell bad, chase the mailman, or roam the neighborhood in packs. Most have very gentle personalities. They present a low public health risk and are not nearly as susceptible to rabies as dogs are. Potbellies generally sleep most of the day away, are quiet, and don't wander far away from the food bowl. Of course, it is up to the owner to be responsible for his pig, to feed and care for his pet properly as well as keep his yard clean.

As with any companion pet, there are issues of over breeding, abuse, and neglect. These issues must be addressed and solutions found to remedy them. Many people who buy potbellies have the wrong expectations in regard to the pig's ultimate size and temperament. These piggies are then discarded. And unscrupulous breeders who mislead the potential customers about the pig's size and weight are a problem. These issues may concern some city governments, and licensing may be put into effect in your city. Limits on the number of pigs that you may own, and spaying and neutering laws are common too.

If your city still has laws against "swine," you may be able to help change them, before getting your pet, by educating them. Many cities realize that potbellied pigs are intended to be pets and are welcomed additions to the neighborhood.

ABOUT THE AUTHOR

Priscilla Valentine has been intrigued with pigs since the age of three when she was gifted with a whimsical, rubber piggy toy. Being a city girl, she was unable to fulfill her ultimate dream . . . to nurture a huge, portly, classic Yorkshire farm pig and spend her days just being with these bright animals. So she went to libraries and read everything she could find about pigs and collected a variety of pig figurines instead.

Ms. Valentine went on to get an English /Education degree at Central Washington University and attended college in London, England.

When the potbellies arrived in the country, Priscilla and her husband Steve eagerly adopted several. She spent so much time with her treasured housepigs that she started training them in order to keep them both from getting bored! She used training to bond with them and to channel their energy into a positive direction.

Training them truly became a passion of Priscilla's, and she won more sanctioned National Advanced Trick Competitions than any other person. Her pigs went on to win the sanctioned World Trick Championship an unprecedented four times.

"Nellie," her most talented pigster, became one of only three Permanent Champions at these feats. Priscilla became avidly involved in the potbellied pig industry and was elected to two terms on the Board of Directors of the National Committees on Potbellied Pigs. She wrote articles for national magazines, helping novice pig owners cope with porcine behavior problems, and she lectured at national pig shows on training.

Priscilla holding Tiffany.

The Valentines purchased more potbellied pigs, then bred and nurtured a newborn litter of them. Priscilla spent her days studying the behavior of this newly introduced, mysterious species. She formulated her own unique training method specially designed to motivate pigs. And she developed a way of socializing baby pigs from birth that made them potentially ideal pets.

The Valentines started taking their pigs out in public to perform at pig shows more and more often. Local admirers wanted to hire the piggies to grace their birthday parties. Then festivals called for the pigs, followed by state fairs and professional sporting events.

The media became enthralled with the energetic, shimmering white, 39-pound "Miss Nellie." She appeared on *Donahue* (twice), the *Today Show, Animal Planet, The Discovery Channel, America's Greatest Pets, Real TV,* Japanese TV, the BBC, and was awarded the $10,000 grand prize in the *America's Funniest People* competition. She's appeared on NBC, ABC, and CBS news multiple times. Nellie appeared in full color in over 20 front-page news articles in just a few years and is a top-draw performer at age eight.

Steve Valentine left his long-term job in order to help Priscilla pursue her dream, performing with pigs full-time. He shares her love and enthusiasm for these stubby-legged, affable creatures.

Nellie and her co-stars, Wilbur and the Duchess of Pork, have made Valentines Performing Pigs the most successful and popular pig performing act in the world. The Valentines and their Oriental companions have performed in front of SRO audiences in virtually every state and are in high demand at large fairs and

half-times at professional sporting events. Their highly-trained pigs have also appeared in national TV commercials and are studio-trained to the nth degree.

Ms. Valentine is probably best known among professional animal trainers for her high training standards. Her performing pigs are extremely energetic, responsive, and quick. Their feats are complicated and often go against a pig's nature, like climbing steep ramps or making multiple high jumps. Her eager pigs are capable of doing a challenging trick in less than five second's time and supercede any other type of animals' capabilities, including dogs.

Priscilla demands 100 percent accuracy from her pigs, and her fast-paced presentation is most often described as "incredible" or "amazing." Even though potbellied pigs in general tend to be slow and non-athletic, the Valentine pigs are so highly motivated that they belie this and perform Olympic-like feats in a flash, on a mere command. This is what separates Priscilla's pigs' performances from other animal presentations.

Her pigs love what they are doing, and it shows by their constantly wagging tails and enthusiasm. Motivation is certainly Ms. Valentine's forte, and she is well-respected for her exceptional talents in the entertainment world.

Priscilla wants her audience members to realize how intelligent these pigs really are and to give them the respect they deserve. She strives to educate as well as entertain. The welfare and future of potbellied pigs as pets are of great concern to her. She feels that if people know what to expect BEFORE they purchase a pig, the chances of the pig being abandoned are slimmer.

Even though Priscilla is now well-known in the animal entertainment field, her situation is unique. She intended to be neither an animal trainer nor an entertainer. She aspired to be a writer and high school English teacher. Because of the time commitment she has with her animals and her desire to make them secure and happy, her eventual career evolved totally unplanned. She learned to deeply understand the nature of pigs through her own personal experience with them as her treasured pets. Never in her wildest dreams did she think she would wind up on a stage, with a microphone in her hand, surrounded by porkers in front of hundreds of people!

Even though Nellie and her bristly pals are now full-time talented performers, they are, most of all, beloved family members to Priscilla and her husband Steve.

Index

C

D

E

F

H

O

P

R

S

T

U

V

W

Y

Z

Valentines Performing Pigs

VALENTINES PERFORMING PIGS WILL MAKE YOUR EVENT A SUCCESS

Have you ever seen a pig "ham dunk" a basketball?
Have you ever seen a pig skateboarding?
Have you ever seen a pig actually spell words?
Have you ever seen a pig "fly" through hoops?

These famous stage porkers perform over 40 incredible feats on stage. They have mesmerized fans at the largest fairs in the U.S. and are always top draw. Valentines Performing Pigs have received the International Association of Fairs and Expositions highest rating possible (5) and it was written in the evaluation, "GREAT family entertainment, good crowds." The pigs are an excellent choice for half-time entertainment at professional sporting events too.

The media adores these mini-pigs, and they are almost sure to get your event front-page coverage. The perky pigs are also studio-trained and have filmed many national TV commercials. The Valentines are members of The International Association of Fairs and Expositions, Western Fairs Association, Washington State Fairs Association, Oregon Fairs Association, and Texas Association of Fairs and Expositions. Their act is fast-paced and adorable.

You will be amazed! For a promo pack, reach us by phone or internet, or contact any national talent agency. "When a pig becomes a ham, the sty's not the limit."

(253) 853-HAMS
valentinesperformingpigs.com
email: NELLIESTAR@aol.com

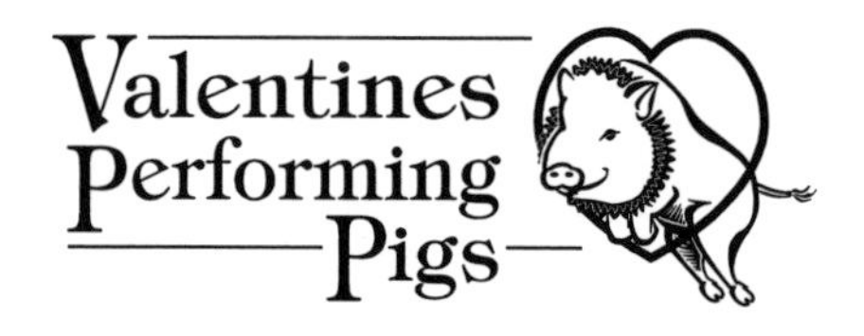